Synthesis *and* Application *of* Lanthanide-Doped Materials

Related titles published by the American Ceramic Society:

Crystal Growth of Novel Electronic Materials (Ceramic Transactions Volume 60)
Edited by R.K. Pandey and Ruyan Guo
© 1995, ISBN 1-57498-003-3

Sol-Gel Science and Technology (Ceramic Transactions Volume 55)
Edited by Edward J.A. Pope, Sumio Sakka, and Lisa Klein
© 1995, ISBN 0-944904-97-1

Science, Technology, and Applications of Colloidal Suspensions (Ceramic Transactions Volume 54)
Edited by James H. Adair, Jon A. Casey, Clive A. Randall, and Sridhar Venigalla
© 1995, ISBN 0-944904-96-3

Materials and Processes for Wireless Communications (Ceramic Transactions Volume 53)
Edited by T. Negas and H. Ling
© 1995, ISBN 0-944904-94-7

Dielectric Ceramics: Processing, Properties, and Applications (Ceramic Transactions, Volume 32)
Edited by K.M. Nair, J.P. Guha, and A. Okamoto
© 1993, ISBN 0-944904-60-2

Experimental Techniques of Glass Science
Edited by Catherine J. Simmons and Osama El-Bayoumi
© 1993, ISBN 0-944904-58-0

Synthesis *and* Application *of* Lanthanide-Doped Materials

▼▲▼

Edited by
B.G. Potter Jr.
Sandia National Laboratories
A.J. Bruce
AT&T Bell Laboratories

▲▼▲

Ceramic Transactions
Volume 67

Published by
The American Ceramic Society
735 Ceramic Place
Westerville, Ohio 43081

Proceedings of the Synthesis and Application of Lanthanide-Doped Ceramics, Crystals, and Glasses Symposium, held at the 97th Annual Meeting of the American Ceramic Society, April 30–May 3, 1995, in Cincinnati.

For information on ordering titles published by the American Ceramic Society, or to request a publications catalog, please call 614-890-4700.

Printed in the United States of America.

1 2 3 4–00 99 98 97 96

ISSN 1042-1122
ISBN 1-57498-012-2

Contents

Optical and Magneto-Optical Applications

Preface

Materials containing lanthanide ions play a critical role in a wide variety of technologies that utilize the optical and/or magneto-optical properties of the dopant. Advances in such diverse fields as telecommunications, information display, radiation detection, and integrated optics rely on the development of improved lanthanide-doped materials that provide enhanced optical performance. Through innovative design and synthesis of the host matrix and hence the local structural environment of the lanthanide ion, the optical response of the dopant population can be influenced. Successful manipulation of the host structure, while maintaining the material properties required for device construction and operation, is essential in order to refine material behavior in existing applications and extend the use of this class of optical materials into new areas requiring a broader range of optical functionality.

This volume contains invited and contributed papers presented at the Synthesis and Application of Lanthanide-Doped Ceramics, Crystals, and Glasses Symposium held at the 97th Annual Meeting of the American Ceramic Society in Cincinnati, April 30–May 3, 1995. The symposium's goal was to provide a forum for the discussion of new and ongoing investigations into the development and behavior of rare earth–doped materials. We wish to thank the speakers, session chairs, and attendees for their contributions to a successful symposium addressing this interesting and technologically relevant subject.

B.G. Potter Jr.
A.J. Bruce

Fundamental Materials Issues

THE ROLE OF LANTHANIDES IN OPTICAL MATERIALS

Marvin J. Weber
Lawrence Berkeley National Laboratory
Berkeley, CA 94720

ABSTRACT

A survey is presented of the use of the lanthanides as chemical components in transmitting optical materials and as activators in materials for luminescent, electro-optic, magneto-optic, and various photosensitive applications.

INTRODUCTION

Lanthanide series elements are used in a multitude of applications—chemical, metallurgical, ceramic, optical, magnetic, and high temperature superconductivity—however their role in optics is one of the most extensive, varied, and significant. Optical applications of the lanthanides span the spectral range from the ultraviolet through the infrared. Their role may be as a constituent of transmitting glasses and crystals, as a colorant, decolorant, or antisolarant, as a luminescent ion in spectroscopy and in laser, phosphor, and scintillator applications, as a magneto-optic ion for optical rotators and isolators, and as an essential element in various photosensitive and photorefractive materials. In applications such as lasers and phosphors, the use of lanthanides is overwhelming compared to most other transition group elements; in other applications the lanthanides play only a minor or negligible role. Lanthanide materials for optics also encompass an impressive range in sizes: from diameters measured in microns for lanthanide-doped optical fibers to dimensions of tens of centimeters for laser amplifer disks and scintillators.

Here we survey the scope and scale of optical applications of materials containing lanthanides. These are principally, but not exclusively, inorganic crystals and glasses. It is not possible to be either inclusive with respect to all applications nor exhaustive with respect to all materials. Instead, a cursory overview of the major uses of lanthanides in optical materials is presented together with a representative range of properties and diversity of chemical compositions. The physical processes involved in their operation are not discussed but are generally well understood.

The discovery and identification of the lanthanides date from 1803 (Ce) to 1945 (Pm). Although the optical spectra of lanthanide ions were used for elemental identification in the late 1800s, the widespread utilization of the lanthanides for optical applications in this century was triggered by and benefited from the concurrence of two developments: (1) an understanding of the optical spectra of the lanthanides derived from many experimental studies and theoretical treatments of crystal fields and of radiative and nonradiative transitions, both conducted primarily in the period from the 1930s through the 1960s, and (2) the availability of high-purity lanthanide materials beginning in the late 1950s-early 1960s.

Lanthanide ions owe their optical properties and extensive utilization to a special combination of spectroscopic features. They are characterized by an electronic structure consisting of an unfilled inner 4f shell and outer filled 5s and 5p shells. When lanthanides are present in a solid, the effects of the crystalline or ligand field are a small perturbation on the 4f states because of the shielding of the outer electron shells. Optical transitions between 4f levels are parity forbidden, hence they are weak and become allowed only by small admixing of states of opposite parity by odd-order terms in the local field or are induced by vibrations. Optical transitions between 4f and 5d states, on the other hand, are electric-dipole allowed and can be intense in solids. Thus the lanthanides exhibit the uncommon (but not unique) feature of having both sharp, weak 4f–4f transitions and broad, intense 4f–5d transitions. The observed optical transitions are predominantly electric-dipole in character (there are, owever a few significant magnetic-dipole transitions). Radiative lifetimes range from about $10^{-8} – 10^{-6}$ s for d–f transitions to $10^{-5} – 10^{-2}$ s for f–f transitions. Competing nonradiative decay by multiphonon emission depends on the vibrational properties of the host and is reduced by selecting hosts with low vibrational frequencies.

Today we have a large choice of available glass and crystal compositions from which to tailor optical and other physical properties of the material for a specific application. As we shall see, the lanthanide ions may be present as a dilute impurity ($< 10^{-3}$ in fiber lasers) or as a stoichiometric component of the material (e.g., in magneto-optic and some laser and scintillator applications).

TRANSMITTING OPTICAL MATERIALS

Of the lanthanide ions, trivalent La, Ce, Gd, Yb, and Lu have large regions of optical transparency in the visible. Materials containing these ions can be used for transmitting optical components such as windows, lenses, and substrates. Glasses are more commonly used for these applications because they are optically isotropic and can usually be produced at relatively low cost and in large sizes. Lanthanides have been incorporated as glass network modifiers into a wide variety of oxide, halide, and chalcogenide glass forming systems (see Ref.1 and the paper by A. J. Bruce and W. H. Grodkiewicz in this volume). These include oxide glasses: silicate, borate, borosilicate, germanate, phosphate, titanate, tellurite, etc.; fluoride glasses: beryllium, aluminum, zirconium, and hafnium fluoride; mixed anion glasses (e.g., fluorosilicates and fluorophosphates); and chalcogenide glasses, such as La_2S_3–Ga_2S_3. Most of these glasses are mechanically and chemically robust and can be produced in sizes and of quality sufficient for optical applications. Less common lanthanide-containing glasses are nitrate glasses[2] $LiGd(NO_3)_4$ and $MgLa(NO_3)_5$ and CH_3- or CF_3-acetate glasses formed using trivalent lanthanides.[3] Refractory lanthanide oxides and Ln_2O_3–Al_2O_3 binaries (where Ln = La, Nd, Sm, Gd, Ho, Er, Yb, Lu) have been produced by more extreme methods such as splat cooling or from molten droplets spun off a laser-heated ceramic rod and solidified by free-fall cooling.[4]

Lanthanum oxide is introduced into silicate, borate, and borosilicate glasses to increase the refractive index. Several commercial optical glasses contain significant quantities of rare earth oxides. These are the lanthanum flint and crown glasses LaF,LaSF and LaK,LaSK, the tantalum flint glasses TaF and TaSF, and the niobium flint glasses NbF and NbSF (Schott designations). These glasses are located in the upper right-hand corner of a refractive index–Abbe number plot of optical glasses shown in Fig. 1. The refractive indices of the simple refractory lanthanide oxide glasses noted above are very high, ~1.9–2.1.[4] Zirconium and hafnium fluoride glasses such as ZBL, ZBLA, ZBLAN, HBL, and HBLA contain small quantities of lanthanum, typically <10 mol. %.

Cubic and uniaxial crystals are also used for transmitting optics. Because of the favorable properties and ready availability of many alkali halide and alkaline fluoride crystals, the use of lanthanide crystals for simple transmitting optical components is rare. Crystals are of more interest for their anisotropic optical properties. The properties of many non-centrosymmetric organic and inorganic crystals have been investigated for electro-optic and nonlinear optical applications. Although the second-order optical coefficients of a few lanthanide crystals [e.g., $K_2La(NO_3)_5$, $Gd_2(MoO_4)_3$, $Nd_xY_{1-x}Al_3(BO_3)_4$] and the third-order optical nonlinearities in isotropic materials (e.g., crystalline $Tb_3Al_5O_{12}$ and SF-7 and LaSF-7 glasses) have been reported,[7] only the neodymium yttrium borate and $LaBGeO_5$:Nd have found any significant use arising from their nonlinear optical properties—as self-doubling laser materials. Crystals used for photorefractive, acousto-optic, and birefringent applications have generally been non-lanthanide-containing materials.

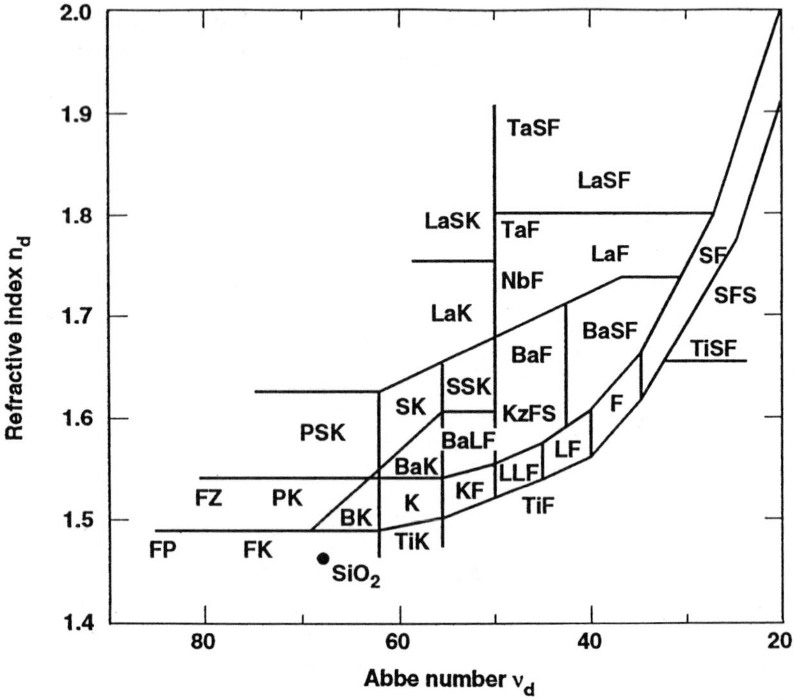

Figure 1. Refractive index vs. Abbe number for commercial optical glasses.

Filter Glass

Filter glasses alter the spectrum of the transmitted optical radiation and thereby provide selective filtering for scientific and technological applications, color enhancement in ophthalmic lenses, and contrast enhancement in displays. The transmission characteristics are determined by a combination of the absorption of the base glass and of dopant ions or colloidal colorants.[5] The ultraviolet (UV) and infrared (IR) cutoff wavelengths vary greatly depending on the glass composition, thus the composition can be tailored to achieve transmission with selected cutoff characteristics. All of the trivalent lanthanides from Ce through Yb (with the exception of radioactive Pm) have been introduced into sodium silicate glasses (in concentrations of ~30 wt. %) for use as filter glasses. Cerium- and gadolinium-containing glasses provide sharp cut-on filters for the near UV; the other lanthanides provide attenuation at a multitude of narrow bands throughout the visible and near-IR regions for special filtering applications.

Colorants and Decolorants

The coloration/decoloration properties of lanthanides in glass have been known and appreciated for many years.[6] When present in large concentrations, lanthanides are colorants in art glass, the most prominent examples being Nd^{3+} for red-violet, Er^{3+} for pink, and Pr^{3+} for green. Lanthanides are also used as decolorants to compensate for coloration due to the presence of unintentional impuirities. For example, divalent iron impurities in oxide materials impart a green coloration whereas trivalent iron has only a pale yellow coloration. Cerium oxide, a strong oxidizing agent, is therefore added to glass as a decolorant to keep iron in the Fe^{3+} state.

LUMINESCENCE MATERIALS

Many forms of luminescence occur in optical materials: photoluminescence (excited by optical radiation), electroluminescence (excited by electrons in an applied electric field), radioluminescence (excited by ionizing radiation), cathodoluminescence (excited by an electron beam), and thermoluminescence (thermally-induced from trapping centers).[8] Photoluminescence from lanthanide series ions in insulating solids is exceedingly common and has led to a multitude of useful applications.

Spectroscopy

The 4f-4f optical transitions of lanthanide ions have narrow homogeneous linewidths. When ions are present in many physically different environments, as in a glass, the transition frequency can serve as a sensitive probe of the local field. If monochromatic laser radiation is used to excite luminescence, only those ions in sites resonant with the laser to within the homogeneous linewidth are excited. This is the basic principle of site-selective spectroscopies such as spectral hole burning and fluorescence line narrowing. Using tunable lasers the entire distribution of sites can be probed. Extensive use has been made of laser-excited luminescence from Eu^{3+} to investigate site-to-site variations in the energy levels and transition probabilities of lanthanides in glasses.[10] As another good example of this technique, laser-excited spectra of Er^{3+} were used to identify 20 different charge compensation schemes for lanthanides substituted at divalent calcium site in CaF_2.[9]

Lasers

Stimulated emission (laser action) requires an inverted ion population of the excited species. The overwhelming number of reported solid-state lasers have been lanthanide ion lasers.[11-13] Trivalent and divalent lanthanides provide a plethora of electronic energy levels and metastable states for optical pumping and low-threshold lasing schemes. Laser action has been reported for all of the lanthanide ions and at wavelengths ranging from the UV to the IR. Host materials have included crystals, glasses, and liquids. Table I lists the many known lasing lanthanide ions, wavelengths, and host materials and is testimony to the widespread use of the lanthanides for solid-state lasers. Neodymium lasers are among the most versatile and widely used solid-state lasers; Er-doped fiber amplifiers are playing an increasingly important role in telecommunications.

Laser action has been demonstrated in hundreds of insulating crystals having many different physical structures and chemical compositions and at wavelengths extending from 0.17 to 7.2 μm.[14] Because the emission lines in crystals are narrower than in glasses, lower lasing thresholds and higher optical gains are possible for crystalline lasers. All trivalent (with the exception of radioactive Pm) and several divalent (Eu, Dy, Tm) lanthanides have been reported to lase in crystals (note: lasing of Gd^{3+} is not well substantiated).

Table I. Lasing Ions, Representative Wavelengths, and Solid-State Laser Hosts.

Ion	Wavelength(s) (μm)	Hosts
Ce^{3+}	0.28–0.32	fluoride crystals
Pr^{3+}	0.49, 0.53, 0.54, 0.60, 0.64, 0.70, 0.72, 0.86, 0.89, 0.91, 1.04, 1.05, 1.3, 3.6, 5.2, 7.2	oxide and fluoride crystals and glasses
Nd^{3+}	0.17, 0.38, 0.93, 1.06, 1.35, 1.8, 5.4	crystals, glasses, liquids
Pm^{3+}	0.93, 1.10	phosphate glass
Sm^{2+}	0.70	fluoride crystals
Sm^{3+}	0.59, 0.65	fluoride crystals
Eu^{3+}	0.61	oxide crystal, liquids
Gd^{3+}	0.31	oxide crystal
Tb^{3+}	0.54	fluoride crystal, liquids
Dy^{2+}	2.36	fluoride crystal
Dy^{3+}	3.0, 4.3	fluoride crystal
Ho^{3+}	0.55, 0.75, 0.98, 1.0, 1.2, 1.4, 1.5, 1.7, 2.1, 2.3, 2.9, 3.4, 3.9	oxide and fluoride crystals and glasses
Er^{3+}	0.47, 0.55, 0.56, 0.67, 0.70, 0.85, 1.0, 1.3, 1.6, 1.7, 2.0, 2.5, 2.8, 4.8	oxide and fluoride crystals and glasses
Tm^{2+}	1.12	fluoride crystal
Tm^{3+}	0.45, 0.48, 1.5, 1.9, 2.3	oxide and fluoride crystals and glasses
Yb^{3+}	1.03	crystals and glasses

An awesome number of lanthanide-containing crystals have been used for lasers. Approximately 175 of the nearly 300 reported crystalline laser hosts contain lanthanide ions are a major cation component. Tables II and III list oxide and halide crystalline hosts (both ordered and disordered) that have been used for lasers. Among the most prominent and commercially available host crystals for trivalent rare earths are $Y_3Al_5O_{12}$, $YAlO_3$, YVO_4, and $LiYF_4$. In so-called stoichiometric lasers, such as PrF_3 and NdP_5O_{14}, the laser ion is a component of the crystal. Of the halide hosts, fluorides have large band gaps and thus are of interest for lasers in the ultraviolet. Examples of UV lasers include the 5d–4f transitions of Nd^{3+} in LaF_3 and Ce^{3+} in $LiLuF_4$. The heavier halide hosts have lower vibrational frequencies, hence extended transmission in the infrared and smaller probabilities for nonradiative decay of excited states by multiphonon emission. Such hosts enable longer wavelength emission, e.g., the 5 and 7 μm lasing of Pr^{3+} in $LaCl_3$.

Why the preponderance and massive use of lanthanides crystals as hosts for insulating solid-state lasers? Because by far the largest number of lasing ions have been the lanthanides, one naturally seek hosts having substitutional sites of corresponding size and valence to accommodate the lasing lanthanide. Trivalent La, Gd, and Lu provide these features and had wide regions of optical transparency. The use of lanthanide hosts for other lasing transition group elements is rare, but has included several gadolinium garnets in which transition metal ions are substituted at an Al or Ga site.

As evident from Table I, many different lasing ions (principally trivalent lanthanides) and host glasses (oxides and fluorides) have been used for glass lasers.[13] The materials have been in the form of rods, disks and slabs, fibers, and planar waveguides. The spectral range of glass lasers extends from ~ 0.5 to 3.5 μm. The spectroscopic properties important for laser action vary with the chemical composition of the glass and can be tailored by varying the chemical composition.[15] Although a large number of experimental laser glasses have been reported in the literature, commercial laser glasses are almost exclusively silicates and phosphates. Phosphate glasses generally have narrower linewidths and, for ultraphosphates, have less fluorescence concentration quenching than for silicate glasses. Large quantities of phosphate glasses doped with Nd^{3+} are used in amplifiers for laser systems which provide the versatility of wavelength and pulse duration needed for present-day inertial confinement fusion experiments.[16] Fluoride glasses, such as ZBLAN, due to their lower vibrational frequencies and hence smaller probability for nonradiative decay by multiphonon precesses, are used as hosts to achieve laser action at longer wavelength lasing.

Table II Crystalline Oxide Laser Materials.[14]

Ln_2O_3, Ln = La, Gd, Er	$Y_3Al_5O_{12}$, Ln = Ho, Er, Tm, Yb, Lu
La_2O_2S	$LnAO_3$, A = Al, Ga, Sc; Ln = La, Nd, Gd, Ho, Er, Tm, Yb, Lu
$LuScO_3$	$Ln_3Sc_2A_3O_{12}$, A = Al, Ga; Ln = Gd, Ho, Er, Lu
ZrO_2–Er_2O_3	$Ln_3(Ln,Ga)_2Ga_3O_{12}$, Ln = La, Lu
Ln_2SiO_5, Ln = Er, Lu	$LnA_2(BO_3)_4$, A = Al, Sc; Ln = La, Nd
$La_2Si_2O_7$	$Ln_3Ga_5DO_{14}$, D = Si, Ge; Ln = La, Nd
$7La_2O_3$–$9SiO_2$	$LnGaGe_2O_7$, Ln = La, Nd, Gd
$LnVO_4$, Ln = Gd, Er	$CaLn_4(SiO_4)_3O$, Ln = La, Gd
$LaNbO_4$	$RLn(MoO_4)_2$, R = Na, K, Rb, Cs; Ln = La, Nd, Gd
$La_2Be_2O_5$	$RLn(WO_4)_2$, R = Na, K, Rb, Cs; Ln = La, Gd, Er, Lu
$LaBGeO_5$	$R_3Nd(PO_4)_2$, R= Na, K
$Ca_4La(PO_4)_3O$	$Na_2Nd_2Pb_6(PO_4)_6Cl_2$
$LaSr_2Ga_{11}O_{20}$	$RLnP_4O_{12}$, R = Li, Na, K; Ln = Pr, Nd
$LuMgAl_{11}O_{19}$	LnP_5O_{14}, Ln = La, Ce, Pr, Nd
$La_2(WO_4)_3$	
$Gd_2(MoO_4)_3$	
$R_5Nd(MoO_4)_4$, R = Na, K	
$NaLnGeO_4$, Ln = Gd, Lu	
$MGdGa_3O_7$, M = Sr, Ba	

Table III. Crystalline Halide Laser Materials.[14]

LnF_3, Ln = La, Ce, Pr, Tb, Ho, Er
$LnCl_3$, Ln = La, Ce, Pr,
$LnBr_3$, Ln = La, Pr
$LiLnF_4$, Ln = Gd, Ho, Er, Yb, Lu
$LiKGdF_5$
α–$NaCaLnF_6$, Ln = Ce, Er
$CsGd_2F_7$
$BaLn_2F_8$, Ln = Er, Tm, Yb, Lu
$K_5NdLi_2F_{10}$
MF_2–LnF_3, M = Cd, Sr, Ba; Ln = La, Ce, Nd, Gd, Ho, Er, Lu
CaF_2–LnF_3–YF_3, Ln = La, Ce, Nd, Gd, Ho, Er
CaF_2–ErF_3–TmF_3–YbF_3
CaF_2–SrF_2–BaF_2–LaF_3–YF_3

As discussed earlier in connection with Fig. 1, glasses containing large quantities of lanthanides are high-index, high-dispersion glasses. The nonlinear refractive index of these glasses is also large. This, however, is detrimental for high power lasers where intensity-dependent changes in refractive index lower the thresholds for beam distortion and optical damage due to self focusing. Fluorophosphate glasses have been developed for laser applications requiring small nonlinear refractive index coefficients to reduce self focusing.

In contrast to the above lanthanide-containing solid-state lasers, alkali halides have been the principal host materials for color center lasers.

Liquid lasers have incorporated lanthanides as the active ion. These have been of two types: (1) chelate lasers where the lanthanide (Eu^{3+}, Tb^{3+}, Nd^{3+}) is part of a complex in an organic solvent, and (2) aprotic lasers consisting of Nd^{3+} and an inorganic complex of a oxyhalide or halide of heavier elements such as phosphorus, sulfur, selenium, or zirconium.[17]

The versatility of lanthanide ions is also demonstrated by their use as sensitizer ions for lasers and other luminescent applications. Many sensitization schemes have been reported in which one lanthanide (the donor) absorbs optical radiation and transfers it radiatively or nonradiatively to another ion (the acceptor).[14] This leads to better utilization of the optical spectrum and an increase in optical pumping efficiency.

The many 4f energy levels of lanthanides also make possible a number of upconversion schemes wherein two or more lower energy photons and transitions are used to populate higher energy levels with resultant emission at energies greater than the pump photons. The development of intense laser diode and laser diode arrays for pumping solid-state lasers in the near infrared has enhanced the prospective use of several upconversion schemes.

Impressive as is the number of demonstrated lanthanide lasers, it is worth noting that for the lanthanide series elements, there are 1639 free-ion energy levels and 192,177 transitions between J states. To date only about 70 transitions have been used for lasing. There is little question that, given an appropriate host and in the absence of excited-state absorption, many more lanthanide lasers are possible.

Phosphors

Many lamp and display phosphor materials are based on the use of lanthanides as the luminescent center. In the past halophosphates activated with luminescent ions other than the lanthanides have been used for lamp phosphors to convert the 254- and 185-nm emission from a low-pressure mercury discharge to visible light. Now, however, tricolor (red, green, blue) lamps containing a combination of lanthanide-activated phosphors provide improved efficiency, lifetime, and color rendering for this application. Examples of these phosphors are included in Table IV.

Probably the most familiar lanthanide phosphor ion is trivalent Eu used for the red color in TV and other CRT displays. Although europium is not an abundant rare earth, its extensive use over the years has resulted in increased availability and an order of magnitude decrease in price. Other cathode-ray tube phosphors have been developed to emit at various wavelengths and with either fast decay for flying-spot scanners or long persistence to avoid or minimize flicker. Commercial lanthanide-activated phosphors for these applications are listed in Table IV.

Table IV. Common Phosphors Containing Lanthanide Activator Ions.

• *Commercial Phosphors*		• *Lamp and Display Phosphors*
P22	$YVO_4:Eu^{3+}$	Red (~610 nm):
P22	$Y_2O_2S:Eu^{3+}$	$Y_2O_3:Eu^{3+}$
P43	$Gd_2O_2S:Tb^{3+}$	$YVO_4:Eu^{3+}$
P44	$La_2O_2S:Tb^{3+}$	$Y_2(WO_4)_3:Eu^{3+}$
P45	$Y_2O_2S:Tb^{3+}$	Green (~550 nm):
P45 (red)	$Y_2O_2S:Tb^{3+}, Eu^{3+}$	$CeMgAl_{11}O_{17}:Tb^{3+}$
P46	$Y_3Al_5O_{12}:Ce^{3+}$	$(Ce,Gd)MgB_5O_{10}:Tb^{3+}$
P46 (green)	$Y_3(Al,Ga)_5O_{12}:Ce^{3+}$	$(Ce,La)PO_4:Tb^{3+}$
P47	$Y_2SiO_5:Ce^{3+}$	Blue (~450 nm):
P48	P46-P47 blend	$BaMgA_{14}O_{23}:Eu^{2+}$
P53	$Y_3(Al,Ga)_5O_{12}:Tb^{3+}$	$Sr_3(PO_4)_5Cl:Eu^{2+}$
P56	$Y_2O_3:Eu^{3+}$	$Sr_2Al_6O_{11}:Eu^{2+}$
	$InBO_3:Eu^{3+}$	
	$InBO_3:Tb^{3+}$	•*Electroluminescent Phosphors*
•*X-ray Storage Phosphors*		Red – $ZnS/CaGa_2S_4:Sm^{2+}$
$BaFBr:Eu^{2+}$		Green – $ZnS:Tb^{3+}$
$Ba_5SiO_4Br_6:Eu^{2+}$		Blue – $CaGa_2S_2:Ce^{3+}$

For projection television, saturation of the light output at high excitation intensities is a problem. Red Eu^{3+} phosphors and green Tb^{3+} phosphors are used but no ideal phosphor has been found for the blue. Phosphors containing Eu^{2+}, Ce^{3+}, and Tm^{3+} have been considered to replace ZnS:Ag^{+} for the blue phosphor in projection TV displays.

Lanthanide-containing ceramic phosphors are used in applications where light is integrated, such as in certain radiation detection applications and in computed tomography (CT). Examples of these materials and their properties are given in Table V. An important consideration for CT applications is the afterglow intensity (due to the presence of traps) which should be small (<0.1% at 100 ms) to preserve image contrast. Of the materials in Table V, $(Y,Gd)_2O_3$:Eu,Pr has a relatively long afterglow (3 ms).

Table V. Characteristics of Ceramic Phosphors for CT Detectors.[8]

Property	Reference CsI:Tl	$(Y,Gd)_2O_3$: Eu,Pr	Gd_2O_2S: Pr,Ce	$Gd_3Ga_5O_{12}$: Cr,Ce
Density (g/cm^3)	4.52	5.91	7.34	7.09
Emission max. (nm)	550	610	510	730
Light output (relative)	100	~67	~75	~60
Decay time (μs)	0.98	~1000	~3	140

Recent phosphor developments have included the discovery of new and improved lanthanide-activated materials for electroluminescent phosphors in flat panel displays, x-ray intensifier screens, and photostimulable storage screens (see Tables IV and V). Although several models have been advanced to describe the processes active in storage phosphors, details of the storage mechanism is still unclear.[18]

Scintillators

Scintillator crystals are widely used for the detection of ionizing radiation in a variety of applications including high energy and nuclear physics, astrophysics, geophysical exploration, medical imaging, security inspection, and industry. Absorption of ionizing radiation leads the creation of large numbers of election-hole pairs, subsequent excitation of the activator, and resultant fluorescence which is detected. Major considerations for the selection of scintillator materials include light yield and associated energy resolution, high density and stopping power to minimize the amount of material required, decay time and afterglow, emission wavelength to match photodetectors, and stability with respect to environmental conditions and radiation damage.[19]

Cerium-activated scintillator materials have been known and used for several decades but have received renewed interest recently because of the favorable spectroscopic properties of Ce^{3+} and the ability to incorporate Ce^{3+} into many different host materials.[20] Among the attractive properties of Ce^{3+} are (1) its luminescence, which occurs in the visible-near ultraviolet region and is well matched to the spectral response of high efficiency photodetectors, (2) the emission involves an allowed transition and thus is fast for good time resolution, (3) the luminescence intensity and decay time are stable and insensitive to temperature for many hosts at room temperature, (4) there are no long-lived radioactive isotopes, and (5) Ce is the least rare of the rare earths. In addition to the 5d-4f transition of Ce^{3+}, similar transitions of Pr^{3+} and Nd^{3+} have been explored for use in scintillators.[21]

Silicate glass was used as a host in early scintillator applications of Ce^{3+}.[22] Scintillation glasses are available in bulk form for energy calorimetry and in fiber form for particle tracking. Cerium-activated glasses containing ^{6}Li, which has a large neutron capture cross section, are also used for detecting neutrons.

The density, dominant decay time, and relative light output of a number of Ce^{3+} activated materials are given in Table VI. All of these materials have emission in the 300 to 500-nm regime which is well matched to the spectral response of photoemissive detector materials. However, there is a large variation in the light yield of Ce^{3+} scintillators; it ranges from values approaching that of the widely used scintillator NaI(Tl) for Lu_2SiO_5 to values several orders of magnitude smaller for some glasses. The large variation in Ce^{3+} scintillation efficiency is due to differences in the efficiency for excitation transfer from the host to the fluorescing activator. Of the materials listed in Table VI, lutetium compounds are attractive because they combine high density and stopping power with high light yield and are the subject of much current research and development.[20]

For applications requiring large volumes of scintillator materials, such as detectors for medical imaging tomography or high-energy physics experiments (where several thousand crystals and volumes of materials measured in cubic meters are needed), production and cost of the scintillator material become major considerations. Scintillation detectors have been proposed for electromagnetic calorimetry at future high-energy particle colliders where about one hundred thousand crystals several centimeters in width and several tens of centimeters in length are needed.[23] Crystals of CeF_3 was seriously considered for this application and would have constituted a massive utilization of a lanthanide, however $PbWO_4$ was selected instead. Although the scintillation properties of $PbWO_4$ crystals were less favorable, their production capabilities were more attractive.

Table VI. Comparison of Properties of Cerium Scintillator Materials.

Material	Density (g/cm^3)	Decay time (ns)	Relative light yield
Reference			
NaI(Tl)	3.7	230	100
Crystals			
Lu_2SiO_5 (LSO)	7.4	~40	75
Gd_2SiO_5 (GSO)	6.7	30–60	20–25
$LuAlO_3$ (LuAP)	8.3	20	~20–50
$YAlO_3$ (YAP)	5.6	28	40
$LuPO_4$	6.5	24	33
CeF_3	6.2	~5, 30	5–9
Glasses			
silicate (GS1)	2.5–2.7	~ 60	~ 5
borate	2.4–2.5	~ 30	~ 3
phosphate	~2.6	~ 30	< 1
fluoride (HfF_4)	~6–7	10–30	≤ 1

MAGNETO-OPTIC MATERIALS

Optical rotation occurs in a medium when the refractive indices n_+ and n_- for right- and left-circularly polarized light are unequal. The microscopic origin of magnetic optical rotation is the inequality that is created by the splitting of the ground or excited-state energy levels due to an applied magnetic field or magnetization. When the magnetic vector is parallel to the direction of light propagation, the Faraday effect produces a rotation of the plane of polarization given by $\theta_F = V\ell H \cdot \varepsilon$, where ℓ is the pathlength in the material, H is the strength of the applied field, ε is a unit vector in the direction of light propagation, and the constant of proportionality V is the Verdet constant. In the small-field limit, the Verdet constant is a function of wavelength, temperature, and refractive index and may be of diamagnetic, paramagnetic, or ferromagnetic origin.[24]

The Faraday effect (which is discussed by J. Kohli elsewhere in this volume) has been utilized in the construction of fast optical switches, modulators, circulators, isolators, and sensors for magnetic fields and electric currents. Diamagnetic and paramagnetic crystals and glasses containing lanthanides have been used in bulk and fiber forms for these applications. The largest diamagnetic Verdet constants are found for high-index, high-dispersion materials such as the oxide glasses located in the upper right-hand corner of the n vs. ν_d diagram in Fig. 1 and chalcogenide glasses. The temperature dependence of V_{dia} is very small which makes diamagnetic materials attractive for high-stability Faraday effect sensors.

Paramagnetic Faraday rotation is dependent on the number of paramagnetic ions present in the material and the frequency of the electronic transitions of the ions contributing to the rotation. The magnitude and the wavelength and temperature dependencies of the paramagnetic Verdet constant vary depending upon the choice of the paramagnetic ion. All transparent materials exhibit diamagnetic Faraday rotation, however when paramagnetic ions are present in sufficient quantities, paramagnetic Faraday rotation may dominate. Because the diamagnetic and paramagnetic effects produce opposite rotations, the sign of the net Verdet constant of a material can change with paramagnetic ion content. The best host materials for paramagnetic rotators are those having a small diamagnetic Verdet constant (low refractive index and dispersion) and capable of incorporating large quantities of the paramagnetic ion. Whereas the concentration of lanthanide ions in glasses is limited by glass formation, lanthanide ions may be a stoichiometric component in crystals and hence produce larger rotations per unit length. Of the paramagnetic lanthanide series ions , Ce^{3+}, Gd^{3+}, Tb^{3+}, and Eu^{2+} have large regions of useful transparency in the visible-near infrared region. Verdet constants of several representative diamagnetic and paramagnetic crystals and glasses are given in Table VII. Terbium is one of the most commonly used ions in paramagnetic rotator materials.

Table VII. Verdet Constants of Selected Lanthanide Materials (633 nm and 300 K).[24]

Crystals	V(rad/T m)	Glasses	V(rad/T m)
LaF_3	2.5	LaSF31	2.6
CeF_3	−114	LaKN12	6.1
EuF_2	−262	ZBLA	3.1
$Dy_3Al_5O_{12}$	−38	cerium phosphate (FR-4)	−31
$Tb_3Al_5O_{12}$	−180	terbium borosilicate (FR-5)	−71

Ferro, antiferro-, and ferrimagnetic materials are characterized by spin ordering even in the absence of an applied field. Non-metallic ferro- and antiferromagnetic lanthanide materials include EuO, EuS, EuSe, EuTe, TmS, and TmSe.[24] Nonmetallic ferrimagnetic materials are used extensively for optical rotation and isolators at wavelengths >1 μm. Of these, yttrium iron garnet ($Y_3Fe_5O_{12}$ or YIG) has been the most widely studied. To enhance its magneto-optic properties, lanthanide ions are substituted either partially or wholly for yttrium. Bismuth substitution greatly increases the absolute value of the saturation Faraday rotation. Examples of materials illustrating these effects are listed in Table VIII.

Table VIII. Faraday Rotation θ_F of Selected Iron Garnets (1064 nm and 300 K).[24]

Material	θ_F (°/cm)	Material	θ_F (°/cm)
$Y_3Fe_5O_{12}$	280	$Pr_3Fe_5O_{12}$	–1789
$Gd_3Fe_5O_{12}$	65	$Nd_3Fe_5O_{12}$	–1130
$Tb_3Fe_5O_{12}$	535	$GdPr_2Fe_5O_{12}$	–790
$Dy_3Fe_5O_{12}$	310	$Gd_2BiFe_5O_{12}$	–3300
$Tm_3Fe_5O_{12}$	115	$Y_2CeFe_5O_{12}$	–22000

Whereas the Faraday effect is a magneto-optic rotation observed in transmission, the Kerr effect is the corresponding effect observed in reflection from metallic materials. Lanthanides are again of interest. CeSb exhibits the largest Kerr effect of any known materials, however because of its a low Curie temperature (16 K) and the large applied fields required to induce the magneto-optic effect, it has not found practical application.

PHOTOSENSITIVE MATERIALS

Photosensitive glasses are a special class of glasses whose optical properties are changed by optical/thermal treatment. Among these are photochromic glasses that change color under ultraviolet or short wavelength light, then revert back to the original state when the light is reduced. The mechanisms involved in photochromism are complex and include treatment of precipitation kinetics and charge carrier diffusion.[25] Photochromism is observed in strongly reduced alkali silicate and other multicomponent silicate and borosilicate glasses, in glasses containing suspensions of crystallites, and in glasses colored by silver halides. An example of a photosensitive glass is lithium silicate containing CeO_2 as an optical sensitizer and silver, gold, or palladium oxide as nucleating agents. This glass photo nucleates when subjected to ultraviolet or ionizing radiation via the process $Ce^{3+} + h\nu \rightarrow Ce^{4+} + e^-$ followed by heating and diffusion of the electron to a nearly silver ion where $Ag^+ + e^- \rightarrow Ag^0$. Upon heat treatment Ag^0 atoms agglomerate to form silver colloids. These are dispersed through the glass and act as nucleation sites for crystal growth which proceeds upon heating at higher temperatures. Because crystallization occurs only in areas exposed to ionizing radiation, spatially selective chemical machining, photoengraving, and photographs are possible.

Light-induced changes in the refractive index and attenuation (photodarkening) have important consequences for the operation of optical fibers and waveguides. These effects have been studied in germanium-doped silica and other inorganic glasses.[24] Glasses containing the lanthanide ions Ce^{3+}, Pr^{3+}, and Eu^{2+} exhibit irreversible changes over a broad spectral range at UV irradiation when subjected to resonant excitation of the 5d state at room temperature. Photodarkening is also observed in Tm^{3+}-aluminosilicate glass due to a nonlinear, multiphoton excitation process.[26]

Cerium is an additive in some photorefractive materials. In the photorefractive effect, two coherent light beams incident on a solid sample produce an interference pattern in the optical intensity. Photoeffects may occur differentially between the light and dark regions. If mobile photocarriers are generated in the light regions, they can diffuse or hop to the dark region where they are trapped and cause changes in the refractive index. Cerium ions substituted at Ba or Sr sites in $Sr_{1-x}Ba_xNb_2O_6$ act as donors to facilitate this photorefractive effect.[27] Cerium is also added to photorefractive $BaTiO_3$ and $LiNbO_3$ crystals.

Most glasses are photosensitive in the negative sense that they become colored (solarize) after exposure to ultraviolet or ionizing radiation. Photodegradation mechanisms are complex and depend on the material composition. To prevent solarization in optically-pumped laser glass and in glass used for UV lamps, small quantities ($\lesssim 1\%$) of polyvalent ions such as cerium, antimony, and molybdenum are added to the glass. Radiation resistant glasses depends on the presence of both Ce^{3+} and Ce^{4+} to capture free electrons and holes.

As a final illustration of a photosensitive lanthanide material, lanthanum-containing zirconium and hafnium fluoride glasses have been used as Brillouin materials for optical phase conjugation.[28]

CONCLUDING REMARKS

As evident from the brief survey above, the role of lanthanide in materials for photonics and opto-electronics has been enormous. These fields remain vigorous areas of research and development. Although the use of the lanthanides have been well exploited, it is not exhausted. One may therefore anticipate further utilization of the special properties of these ions in optical materials.

ACKNOWLEDGMENTS

This work was supported in part by the Director, Office of Energy Research, Office of Health and Environmental Research, Medical Applications and Biophysical Research Division of the U.S. Department of Energy under contract No. DE-AC03-76SF000098, and in part by Public Health Service Grant R01 CA48002 awarded by the National Cancer Institute, Department of Health and Human Services.

REFERENCES

1. N. J. Kreidl, in *Glass: Science and Technology*, Vol. 1, D. R. Uhlmann and N. J. Kreidl (Eds.), (Academic Press, New York, 1983), p. 105-299.
2. L. E. Topol and R. A. Happe, *J. Non-Cryst. Solids* **15**, 116-124 (1974).
3. L. G. van Uitert and W. H. Grodkiewicz, Mat. Res. Bull. **6**, 283-292 (1971).
4. L. G. van Uitert, W. A. Bonner, and W. H. Grodkiewicz, Mat. Res. Bull. **6**, 513-518 (1971).
5. L. M.Cook and S. E. Stokowski, in *Handbook of Laser Science and Technology , Vol. IV: Optical Materials,* (CRC Press, Boca Raton, 1986), p. 93-183.
6. See, for example, an early report by N. J. Kreidl, *J. Am. Ceram. Soc.* **25**, 141-143 (1942).
7. M. J. Weber (Ed.), *Handbook of Laser Science and Technology, Supplement 2: Optical Materials,* (CRC Press, Boca Raton, 1995).
8. G. Blasse and B. C. Grabmeier, *Luminescent Materials* (Springer-Verlag, Berlin, 1994).
9. D. R. Tallant and J. C. Wright, *J. Chem. Phys.* **63**, 2074 (1975).
10. See, for example, M. J. Weber, *Ceram. Bull.* **64**, 1439-1443 (1985).
11. A. A. Kaminskii, *Laser Crystals*, Second Edition, (Springer-Verlag, Berlin, 1990).
12. M. J. Weber (Ed.), *Handbook of Laser Science and Technology, Vol. I, Lasers and Masers* (CRC Press, Boca Raton, 1981).
13. M. J. Weber (Ed.), *Handbook of Laser Science and Technology, Supplement 1: Lasers*, (CRC Press, Boca Raton, 1991).
14. A. A.Kaminskii, *Crystalline Lasers: Physical Processes and Operating Schemes* (CRC Press, Boca Raton, in press).
15. M. J. Weber, *J. Non-Cryst. Solids* **123**, 208-220 (1991).
16. J. E. Marion and M. J. Weber, Eur. J. Solid State and Inorg. Chem. **28**, 271-287 (1991).
17. H. Samelson, in *Handbook of Laser Science and Technology, Vol. I, Lasers and Masers*, (CRC Press, Boca Raton, 1981), p. 397-422.
18. G. Blasse, *J. Alloys and Compounds* **192**, 17-21 (1993).
19. G. F. Knoll, *Radiation Detection and Measurement*, 2nd edition (Wiley, New York, 1989), p. 215-250.
20. C. W. E. van Eijk, J. Andriessen, P. Dorenbos, and R. Vissor, *Nucl. Instr. and Meth.* **A348**, 546 (1994).
21. C. Dujardin, C. Pedrini, D. Bouttet, et al., IEEE Trans. Nucl. Sci. (in press).
22. R. J. Ginther and J. H. Shulman, *IRE Trans. Nucl. Sci.* **NS-5**, 92-95 (1958).
23. P. Lecoq, *J. Lumin.* **60 & 61**, 948-955 (1994).

24. M. N. Deeter, G. W. Day, and A. H. Rose, in *Handbook of Laser Science and Technology, Supplement 2: Optical Materials,* (CRC Press, Boca Raton, 1995), P. 367-411.
25. R. J. Araujo, in *Treatise on Materials Science and Technology*, Vol. 12, M. Tomozawa and R. H. Doremus (Eds), (Academic Press, New York, 1977) p. 91-122.
26. M. M. Broer, SPIE Proc. Vol. **2044**, 316 (1993) and references cited therein.
27. R. C. Baetzold, Phys. Rev. B **48**, 5789 (1993).
28. L.-G Hwa, et al., *J. Opt. Soc.Am. B*, Topical meeting on Nonlinear Optical Properties of Materials, 833-839 (1989).

LANTHANIDE CONTAINING OXIDE GLASSES.

A.J. Bruce and W.H. Grodkiewicz.
AT&T Bell Laboratories, Murray Hill, NJ 07974, USA.

ABSTRACT.

Lanthanide oxides are considered in the role of major glass constituents (>5 mol%) in oxide glasses. Binary, and/or ternary, silicate, borate are phosphate glasses, in which no alkali or alkaline earth modifiers are present are used as examples to illustrate the effects of the lanthanide ions on glass structure and properties. It is evident that the lanthanide ions behave primarily as glass modifiers. However, unlike more common modifiers these ions promote exceptional stability in the glasses through strong ionic interactions.

INTRODUCTION.

Lanthanides have been used as optical dopants in glass for many years. They are key to many technologically important devices including lasers and optical amplifiers. However, comparatively few authors [1-5] have considered lanthanides in the role of major glass constituents (>5 mol%). Our current level of understanding of such materials is well documented in a recent book edited by Shelby [5]. Potential applications for these glasses are largely undefined, although they have been shown to be promising for Faraday rotation of polarized light [6] and as biomedical radiation delivery vehicles [7]. From a glass formation, structure and properties stand point they are also intriguing.

In this paper we will use examples of binary and/or ternary silicate, borate and phosphate glasses to illustrate glass formation and the structure and property characteristics of these glasses. Alkali and alkaline-earth free glasses have been selected to identify the role of the lanthanides unambiguously as glass modifiers. Glass formation is considered with respect to conventional bulk melting procedures. Yttrium exhibits similar behavior and is also considered as a "lanthanide" in this work.

SILICATE GLASSES.

Binary lanthanide silicate glasses with more than 0.5 mol % lanthanide oxide cannot be prepared without inhomogeneity. Ternary silicate glasses can be formed, including lanthanide alumino-silicates used herein as examples. The extent of glass formation for some different lanthanides is shown in figure l. (after White and Day, reference 5). This glass formation is quite extensive and centers roughly on 20 Ln_2O_3-20 Al_2O_3-60 SiO_2 compositions (in mol%, where Ln signifies a lanthanide.)

In our work we have prepared glasses with this central composition for the Lanthanide series (excluding Pm) and Y. Batch sizes up to 1000g were prepared by conventional melting up to 1550°C. in Pt crucibles and cooled in the furnace. The melts were typified by a low viscosity and somewhat more rapid quenching was required to achieve glass formation for batches containing elements near the beginning of the lanthanide series (La, Ce, Pr, Nd and Sm).

IR reflectance measurements indicate the presence of non-bridging oxides (NBOs) in such glasses [4]. This is consistent with the lanthanides acting as modifiers in a conventional silica network. Based on valence, each lanthanide can contribute three NBOs to the network. Even with a reduction in NBOs due to alumina incorporation (1 per Al) these glasses will remain highly modified.

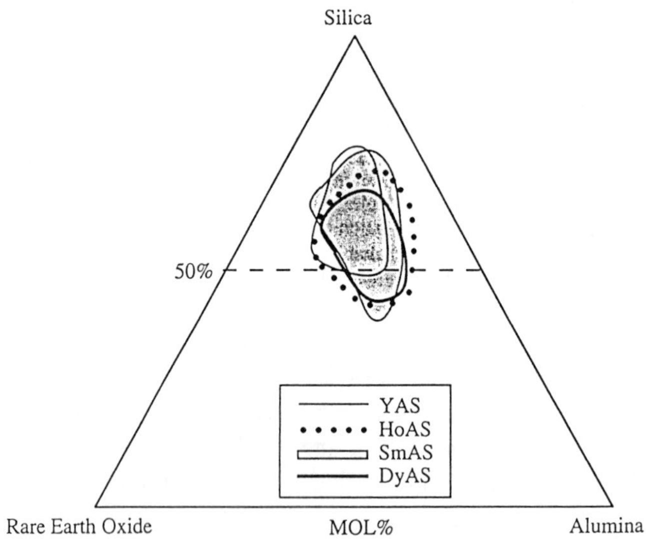

Figure 1. Bulk glass formation for Lanthanide Alumino-Silicate compositions

The lanthanide alumino-silicate glasses are typified by a high resistance to attack by moisture, but not to acid. They exhibit edifying property trends as a function of the size of the lanthanide ion (this decreases along the series as a consequence of the Lanthanide contraction effect). As shown in figure 2 the thermal expansion coefficient (tec) increases, while the glass transition (T_g) and dilatometric softening points (T_d) decrease, with ionic size. Vickers microhardness (H_v) also decreases and the refractive index (n_d) increases with ionic size.[5].

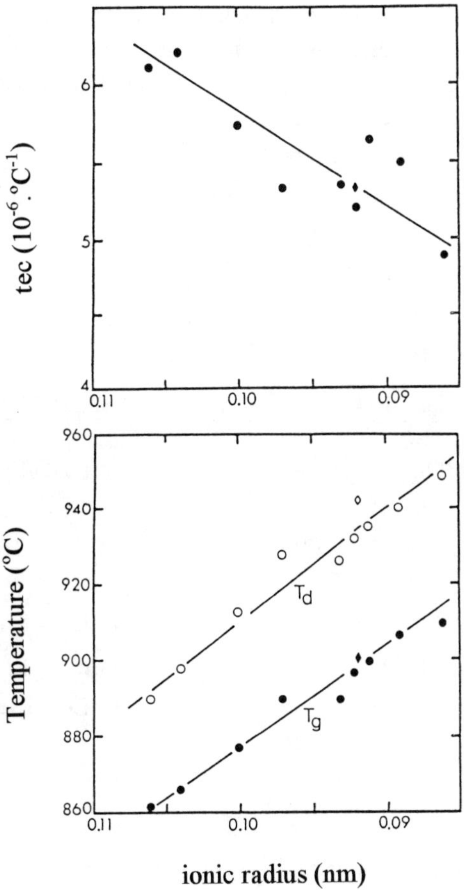

Figure 2. Some physical property trends of 20 Ln_2O_3-20 Al_2O_3-60 SiO_2 glasses.

It is illustrative to contrast the median properties of our lanthanide alumino-silicate glasses with a more familiar sodium di-silicate composition as shown in Table 1. Of the two, the alumino-silicate will have a significantly higher fraction of NBOs/Si and presumably a lower degree of covalent network connectivity. At first appraisal it is therefore unusual that such a glass should exhibit a high glass transition temperature, modest thermal expansion coefficient, high hardness and exceptional durability in water. The rational is that, although more fragmented, the silica network is held together through strong mainly ionic interactions with the lanthanide ions. By virtue of their small size and high valence these will posses a high field strength and participate in strong associations particularly with Nabobs. The trends in physical properties noted above are consistent with changes in field strength as a function of ionic size. Also the rapid decrease in viscosity with temperature above the glass transition is consistent breaking ionic linkages. In the terminology of Angel and co-workers these would be "fragile" glasses.

Such glasses should be considered as candidates for applications in which high temperature, modest expansion, high index and/or excellent durability, if not low cost, materials are required.

Table 1. Silicate glass properties

20 Ln_2O_3-20 Al_2O_3-60 SiO_2	33 Na_2O-67 SiO_2
• NBO/Si = 1.33.	• NBO/Si = 1.0
• T_g = 880°C.	• T_g = 435°C.
• tec = 55. 10^{-7}/°C	• tec = 160 . 10^{-7}/°C
• p = 4.4 g/cm^3	• p = 2.46 g/cm^3.
• H_v = 7.2 GPa.	• H_v = 3.9 GPa.
• n_d = 1.69	• n_d = 1.505.
• High Durability.	• Low Durability .

BORATE GLASSES

Binary lanthanide borates can be prepared in extremely narrow compositional ranges close to 25 Ln_2O_3 -75 B_2O_3 as illustrated in figure 3 for a Neodymium glass [5]. In our own work, we found that we could prepare pure binary glasses with La through Sm, mixed lanthanides of Eu, Gd and Tb as $La_{0.25}$, $Ln_{0.75}$ glasses and the remainder of the series with $La_{0.5}$, $Ln_{0.5}$ mixtures. It is notable that this is the reverse trend of glass formation vis a vis the lanthanide series from the previously mentioned alumino-silicates. In the Borate system the larger lanthanides are obviously more compatible with glass formation.

We prepared our glasses by mixing lanthanide oxides with an excess (5x) of B_2O_3. After melting at 1300°C for 12 hours in Pt crucibles and quenching, a two layered product was obtained. The upper layer was B_2O_3 which could be washed away in water leaving a lanthanide borate glass. Because the interface was somewhat irregular in contour, this residue was usually remelted for 30 minutes at 1300°C with a small addition of B_2O_3 to compensate for volatilization. Without such additions glasses were not obtained on remelting. While the initial glass is presumably saturated with B_2O_3 a minor loss of B_2O_3 on remelting is clearly sufficient to move outside the glass forming range. These are extremely narrow.

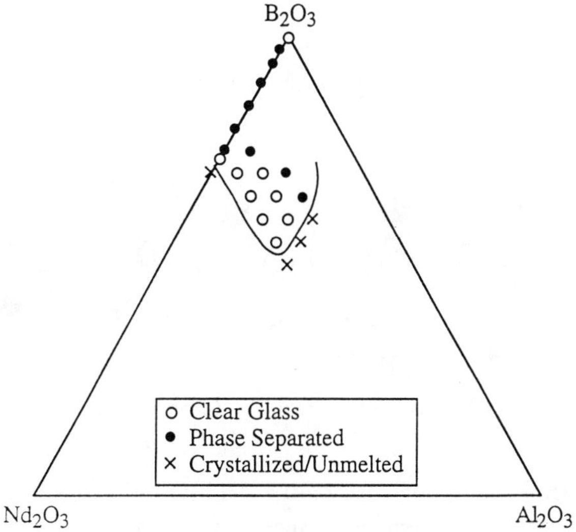

Figure 3. Bulk glass formation in the Nd_2O_3-Al_2O_3-B_2O_3 system

Based on Raman and IR transmission spectra we surmise that the lanthanides act as modifiers in these borate glasses. The IR transmission spectra of a $25La_2O_3$-$75B_2O_3$ glass and crystal of the same stoichiometry are shown in figure 4. Their similarity has prompted the suggestion [8] that the glass has a structure close to that of the crystal (also shown in figure 4).

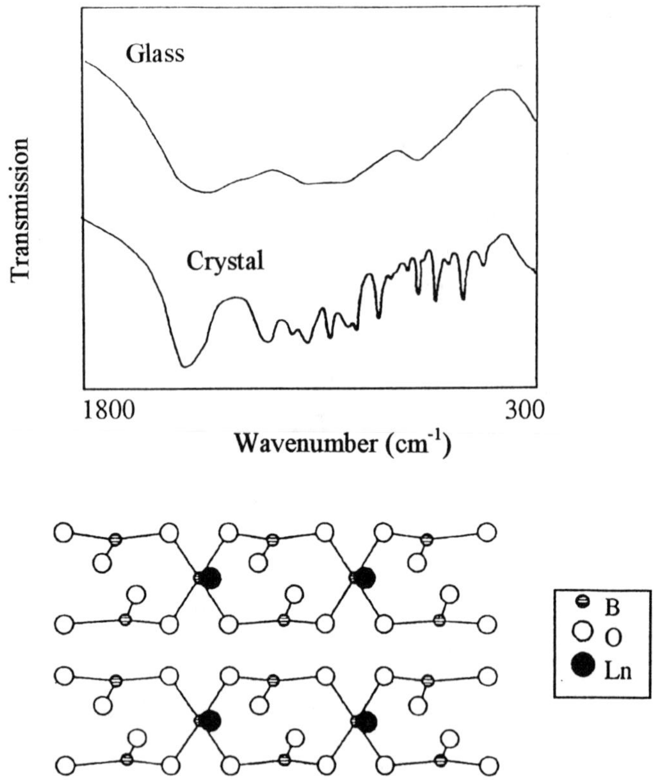

Figure 4. IR transmission spectra of a $25La_2O_3$-$75B_2O_3$ glass and crystal. Also the proposed structure of the same crystal.

The lanthanide borates show similar trends in their properties vis a vis the lanthanide series, to the alumino-silicates considered above. Thermal expansion coefficients and refractive indices increase while glass transition temperatures and Vickers microhardness decrease with increasing ionic radii. These glasses have the highest durability, in water, of any borates previously investigated by the authors.

A comparison of properties is offered in Table 2 between median values for $25Ln_2O_3$-$75B_2O_3$ glasses and a $33Na_2O$-$67B_2O_3$ glass. As with the alumino-silicates, lanthanide borates are typified by high T_g, modest tec, high H_v, high refractive indices and high durability. The stability of the glasses is again consistent with strong, multiple ionic associations between the lanthanides and the borate lattice. However the narrowness of their glass forming range is extremely unusual and should be investigated further.

Table 2. Borate glass properties.

25 Ln_2O_3-75 B_2O_3	33 Na_2O - 67 B_2O_3
• T_g = 720°C.	• T_g = 460°C.
• tec = 70.10^{-7}/°C	• tec = 112. 10^{-7}/°C
• p = 4.0 g/cm³.	• p = 2.36 g/cm³.
• H_v = 7.0 GPa.	• H_v = 4.5 GPa.
• n_d = 1.72	• n_d = 1.517.
• High Durability.	• Low Durability .

PHOSPHATE GLASSES.

There is a continuous range of glass formation in binary lanthanide phosphate glasses from 0 to approximately 30 mol% Ln_2O_3 as illustrated in figure 5 [9]. In this instance we did not prepare a complete series of lanthanide glasses but specific compositions for optical investigations. Glass preparation was by conventional melting of an oxide batch in Pt crucibles at 1300°C for 1 hour, followed by quenching in crucible and annealing. For phosphate glasses these compositions are exceptionally durable.

Raman spectra are compared in figure 6 for La_2O_3-$5P_2O_5$ and sodium metaphosphate glasses. The spectra are very similar in form and the strong peaks associated with non bridging oxides support are consistent with the role of lanthanides as network modifiers. The intensity and position of the spectral peaks for the lanthanide glass is also consistent with stronger associations in this glass.

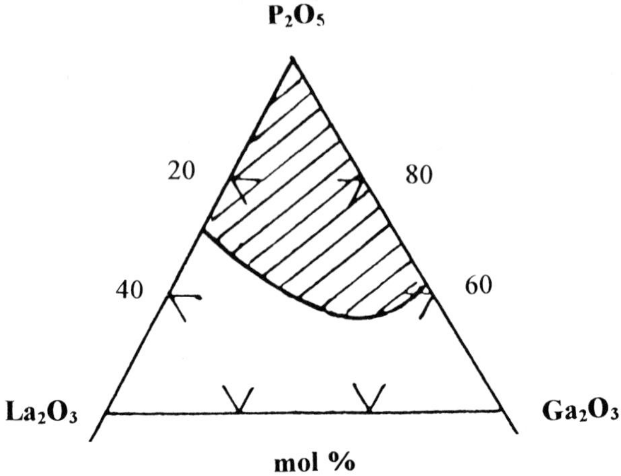

Figure 5. Bulk glass formation (shaded area) in the La_2O_3-Ga_2O_3-P_2O_5 system

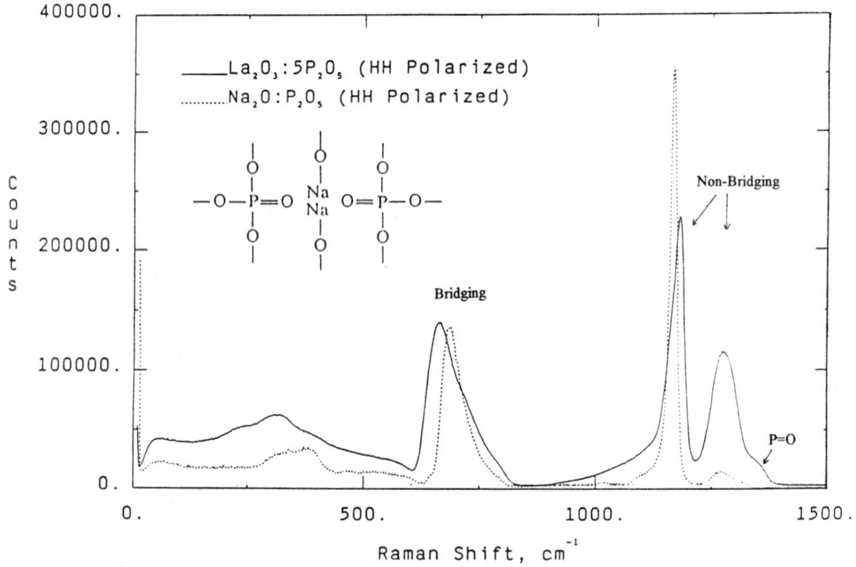

Figure 6. Raman spectra of La_2O_3-$5P_2O_5$ and $NaPO_3$ glasses.

As far as they were determined, the trends in glass properties as a function of the lanthanide series followed those for the silicate and borate glasses noted above. A comparison of typical properties is offered in Table 3 for Lanthanide and sodium metaphosphate glasses. These compositions should have identical fractions of NBO/P. Once more the contrasts are substantial.

Table 3. Phosphate glass properties.

$Ln(PO_3)_3$	$NaPO_3$
• NBO/P = 2	• NBO/P = 2
• T_g = 700°C.	• T_g = 280°C.
• tec = 65. 10^{-7}/°C	• tec = 170. 10^{-7}/°C
• p = 3.4 g/cm^3	• p = 2.48 g/cm^3.
• H_v = 3.2 GPa.	• H_v = 1.5 GPa.
• n_d = 1.57	• n_d = 1.485.
• Fair Durability.	• Low Durability .

SUMMARY

In this paper we have sought to highlight some of the properties of oxide glasses with high lanthanide contents. We have used as examples compositionally simple silicate, borate and phosphate glasses to minimize masking effects from other glass constituents. Without exception the glasses showed remarkable properties for each class of network former. They are tough, durable, have lower expansion and higher transition temperatures than similar glasses with more common glass modifiers.

All the evidence identifies the lanthanides in the role of network modifiers. The properties of the glass would seem to reflect multiple, high strength, ionic interactions between the lanthanides and the dissociated glass networks. When the potential energy of these interactions is surmounted, above T_g, a rapid decrease in viscosity can be observed emphasizing the ionic nature of these materials.

Few applications have been considered for these materials. It may be quite rewarding to consider applications which can exploit their ionic nature (e.g. ionic transport) and physical properties which are often similar to those of highly covalent network glasses.

ACKNOWLEDGMENTS.

The authors wish to thank Drs. R. K. Brow and D. Tallant of Sandia National Laboratories, Albuquerque, NM. for measuring and analyzing Raman spectra for several of our borate and phosphate glass samples.

REFERENCES.

1. J. Loffler, Glatechnische Berichte, 10, 204, 1932.
2. A. Makishima, M. Kobayashi and T. Shimohiri, J. Am. Ceram. Soc., 61, 247, 1978.
3. M.J. Hyatt and D.E. Day, J. Am. Ceram. Soc., 70, C283, 1987.
4. A.J. Bruce and W.H. Grodkiewicz, Paper 97-G-89, Ann. Meeting Am. Ceram. Soc., Indianapolis, IN., 1989.
5. Rare Elements in Glasses, J.E. Shelby, ed., Trans Tech Publications, Aedermannsdorf, Switzerland, 1994.
6. J.T. Kohli, this volume.
7. E.M. Erbe and D.E. Day, J. Am. Ceram. Soc., 73, 2708, 1990.
8. I.N. Chakraborty and D.E. Day, J. Am. Ceram. Soc., 68, 641, 1985.
9. Z.N. Shecheglova, in Handbook of Glass Data, eds. O.V. Mazurin, M.V. Streltsina and T.P. Shvaiko-Shvaikovskaya, p 750, Elsevier,1991

Emissive Materials and Active Waveguides

EVOLUTIONARY DEVELOPMENTS IN LASER GLASS

J D Myers, Founder, Kigre, Inc., 100 Marshland Road, Hilton Head, SC 29926.

Abstract

The history of laser glasses is reported with reference to both neodymium and erbium laser glasses. In addition a new holmium laser glass is reported, which operates at room temperatures. Diode pumping of laser glasses is discussed and results are given which open some interesting areas for further research and development.

Neodymium Laser Glass

In order to discuss the developments made, an overview of the initial performance parameters are given below in Table 1. Here the starting point and current state of the art are compared, over a quarter century span [1,2].

Table 1. Performance Parameters of Neodymium Laser Glass

Performance Parameter	1970	1995
Glass Name	MG-915	QX\Nd
Glass Type	Silicate	Phosphate
Threshold (1/4 x 3" Rod)	20 J	5J
Power output (3/8 x 6" Rod)	5 W	105W
Damage Threshold (30 ns)	40 J	40J
Thermal Behavior	Wild	Mildly athermal

In 1970, the laser glass of the moment was called MG-915, and was made in France by Sovirel for use in the CGE Ud 320C Laser System.....with the following (astounding at the time) output specifications:

Output Energy Per Pulse..65 Joules
Pulse Width..80 nanoseconds
Beam Divergence....................................... ....180 microradians

The USA at that time was thought not to have a comparable laser glass, and US manufacturers were brought to task, and to Washington DC, to explain this unforgivable lapse in technology. In addition, experts from government and university laboratories were assembled to examine and advise. Of all the theories advanced, a favorite was the one that attributed the superiority of the French glass to a proprietary ingredient not available to US manufacturers. That ingredient was identified as UNOBTAINIUM. As we all now know the magic ingredient was the lack of platinum particles in the glass. MG-915 was a Barium Crown Silicate with a 700 microsecond lifetime, a cross section of 1.2 x 10-20, and a loss coefficient of 0.07%/CM at one micron. (The long lifetime was thought necessary to accommodate flashlamp characteristics.) In other words, MG-915 was a good optical glass, but relatively inefficient as a laser glass, compared with what was to come.

QX is a new laser glass...the culmination of research and development efforts extending over a period of years [3]. The objective of this effort was to increase by an order of magnitude, the average power capability of laser glass. In other words, instead of being limited to 1 Hertz, we wanted to be able to operate glass lasers at 10-20 Hertz, without going to exotic engineering schemes like moving slabs or Gatling-Gun configurations. Prior to the development of QX Glass, the maximum average power that one could achieve from a 3/8 x 6 inch glass laser rod was limited to approximately 10 watts. Above this point, the glass rod would either break or the output would deteriorate due to thermally-induced distortions. As indicated on the graph, we at Kigre, Inc., have operated a 3/8 x 6 inch QX/Nd Rod in the laboratory at average powers in excess of 100 watts...with no engineering tricks. This feat represents a major improvement in laser glass performance.

How this new glass was developed is a good example of traditional glass composition research...and a good example of the frustration and intuition that accompany such research. The combination of properties a laser glass must have to provide the order-of-magnitude improvement in output power was seen as follows:

1. The glass must "lase" efficiently...therefore, we have to maintain a high cross section.

2. The glass must not suffer greatly from thermal distortions during operation...therefore, a low thermo-optic constant must be retained, thermal and stress-induced optical distortions must also be reduced.
3. The glass must not break under moderate to high thermal loading...therefore, the thermal expansion must be reduced and/or the rupture strength increased.
4. The glass must not dissolve in water, therefore, durability must be maintained or increased.

Conventional wisdom might say, "Easy, go to a borosilicate, or to fused quartz", the problem is that borosilicates have a cross section of 1.1 and quartz has a cross section of 1.4...making both relatively poor candidates. Previous work on the lasing properties of different glass compositions, [4] (sponsored and beautifully tabulated by LLNL) reveal the composition changes which affect particular laser glass properties. This and other research has provided a glossary of remedies for each and every one of the four desired changes or retentions mentioned previously. For example:

TO INCREASE EFFICIENCY...go to the heavier alkali metals and use more of them. HOWEVER...greater use of alkalis also results in greater thermal distortions, lowers durability, and reduces rupture strength.
TO INCREASE THERMAL SHOCK RESISTANCE...add / increase alumina content. HOWEVER...increased alumina or other R203 will decrease efficiency and usually increase thermal distortions.

In any case, these are static considerations for the most part, and here is envisioned a thermally dynamic environment best described by the thermal-optic coefficients, W, P, and Q. Where: W designates the change in optical path with increasing temperature...the conventional thermo-optic coefficient; P designates the optical path difference due to the co-existence of temperature gradient and stress; and Q designates the stress-induced optical path difference.

$$W = B + a (n - 1)$$

$$P = B - aE (C_1 - 3C_2) \backslash 2 (1 - u)$$

$$Q = aE(C_1 - C_2) \backslash 2 (1 - u)$$

where; $B = dn\backslash dt = C_1 - C_2$, a = coefficient of thermal expansion, n = index of refraction, E = Young's Modulus ($E = 5.3 \times 10^{10}$ Pa), u = Poisson's ratio (u = 0.271), and, C_1 & C_2 are stress-optical coefficients, dependent on piezo-optic coefs.

For a complete discussion of W, P, and Q, see V.C. Kravchenko and Yu. P. Rudnitskii [5]. According to John McMahon of Naval Research Laboratory [6], athermalization occurs at some temperature (T) where [P] = Q/2, given that both P and Q are small.

Thus, the task was to design a laser glass whose operating temperature under multi-pulse operations was such as to generate the conditions for athermal behavior. Previous experiments, showed how to make a laser glass which was both efficient and reasonably athermal under multi-pulse operation, the challenge was to modify a version of this glass to extend its operating power by an order of magnitude. The procedure for accomplishing this task was composed of many steps of refinement and improvement...trial and error if you will. Typically, a melt is made one day, and the resultant laser rod from that melt would be tested two days later, using planar mirrors. This method of evaluation automatically defined an acceptable region of athermal behavior at a realistic operating temperature. Needless to say, "modifications" ended up as entirely new compositions. Through a judicious choice of constituents, our labs managed to devise a low-expansion phosphate glass [3] which has the chemical durability of a silicate, the cross section of a phosphate and can be ion-exchanged to increase the rupture strength without inducing destructive internal reflections. The performance of QX glass doped with neodymium (QX/Nd) is given relative to standard phosphate glass (Q98) in Figure 1.

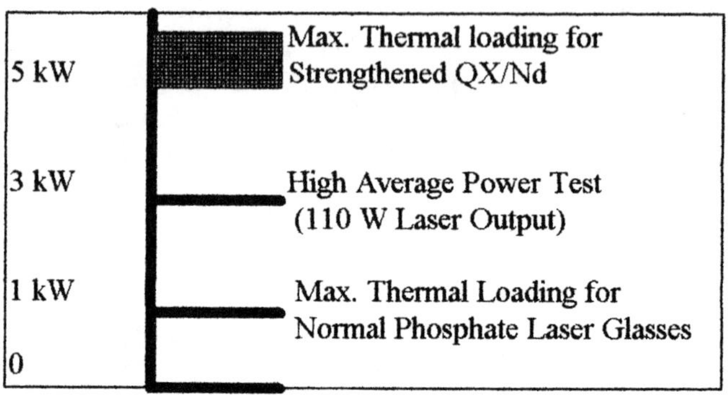

Figure 1 Thermal Loading Capability of Strengthened QX/Nd Glass

Figure 1 relates the breakage point of a 10 x 160 mm QX/Nd rod in a laser pump cavity versus the power input into the flashlamp. A conventional phosphate laser rod of the same size will break at approximately one kilowatt of input power. Strengthened QX/Nd survives up to five kilowatts of input power. Operating at the three kilowatt level, the superior thermo-optic characteristics of this material permitted successful laser operation at a comfortable 3/5 of its rupture strength.

The power input/output curve of the QX/Nd rod is given in Figure 2 [7]. Here the shaded portions depict the breakage points of conventional phosphate glass laser rods and unstrengthened QX/Nd. It should be noted that unstrengthened QX/Nd survived input powers in excess of 300 watts per inch of pump length. Our old friend, ED-2, a silicate, under similar conditions, would not survive input powers of 200 watts per inch of pump length.

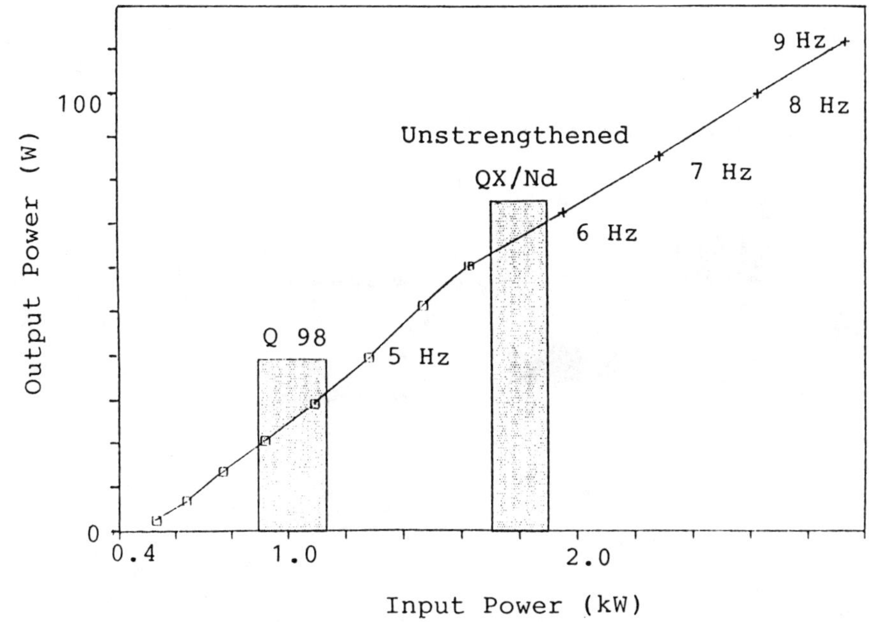

Figure 2 Laser Output vs. Pump power for Strengthened QX\Nd Glass

Erbium Laser Glass

Having developed the QX base, the next step was to doped it with erbium, ytterbium and chromium for flashlamp pumped operation at the "eye-safe" wavelength of 1.54 microns. The BRH standards for Class I lasers are given below in Table 2. Radiation at 1.54 microns is afforded a factor of 100 times more single pulse energy than other wavelengths[8].

Table 2. BRH Standards For Class I Lasers

Wavelength (nm)	Laser	Energy/Shot (mJ)	Average Power (mJ)
690	Ruby	0.0002	0.0004
900	GaAs	0.0005	0.095
1060	Nd:YAG	0.005	0.195
1540	Er:Glass	7.9	0.79
1660	Er:YAP	0.079	0.79
2060	Ho:YLF	0.079	0.79

An energy level diagram illustrating the sensitization pathway is given in Figure 3. Here the chromium acts as a sensitizer to the ytterbium, which in turn acts as a sensitizer to the erbium. Obviously, this scheme would never work if it were not for the 8 millisecond lifetime of erbium I13/2 transition.

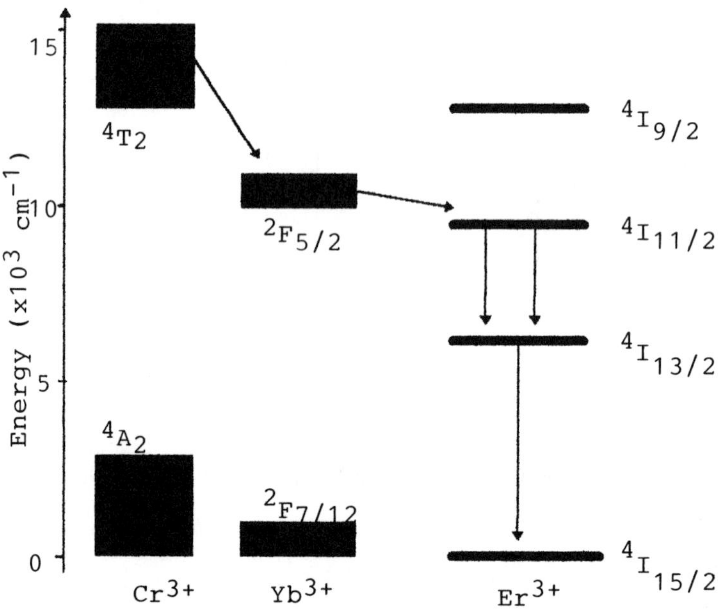

Figure 3 Energy level diagram for Cr^{3+}, Yb^{3+}, Er^{3+}

Due to the addition of the sensitizer chromium to the material, QX/Er as it is called, is a significantly greater absorber of flashlamp energy than its neodymium cousin. Over the same 25 year period, erbium laser glasses have improved by a factor of 75 in pulse repetition rate and average power. Table 3 illustrates the salient features of ED-7, an Owens-Illinois silicate composition similar to ED-2; QE-7S, a Kigre chromium-sensitized phosphate [9]; and QX/Er, another Kigre chromium-sensitized phosphate [3].

Table 3. Advances in Laser Glass ---- Erbium

Performance Parameter	1970	1988	1995 [10]
Glass Name	ED-7	QE-7S	QX/Er
Glass Type	Silicate	Phosphate	Phosphate
Threshold (3 x 50mm rod)	50 J	9 J	9 J
Slope Efficiency	0.2%	2.2%	2.2%
Thermal Behavior	Wild	Athermal	Mildly Athermal
Repetition Rate	0.2 Hz	3 Hz	15 Hz

Figure 4 illustrates the thermal loading capability of strengthened QX/Er:Glass, the limits are similar to those observed in its neodymium counterpart. Tests here were run on a 3 inch rod, resulting in lower total wattage loading than in the neodymium material tests. The erbium-chromium doped material is more highly absorbent than the neodymium version, resulting in a reduction of the thermal loading capacity during flashlamp operation by approximately 60%. Even so, QX/Er with a maximum pump limit of 500 watts per inch, was successfully operated, with liquid cooling, at a thermal loading of over 300 watts per inch.

Figure 5 shows the laser output vs. pump power for a strengthened 3 mm x 75 mm QX/Er laser rod. Here the rod is operated at 10 Hz, over a wide input range. The relatively non-linear portion of the curve between 300 and 400 watts is attributed to the onset of a certain amount of thermal lensing within the resonator. Here the limits of conventional QE-7S material, and unstrengthened QX/Er is shown by shaded areas.

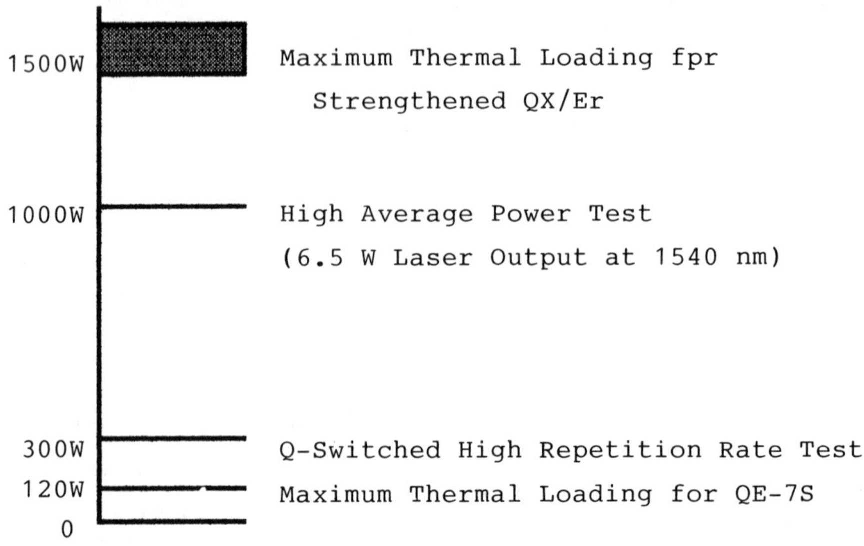

Figure 4 Thermal Loading Capability of Strengthened QX/Er Glass

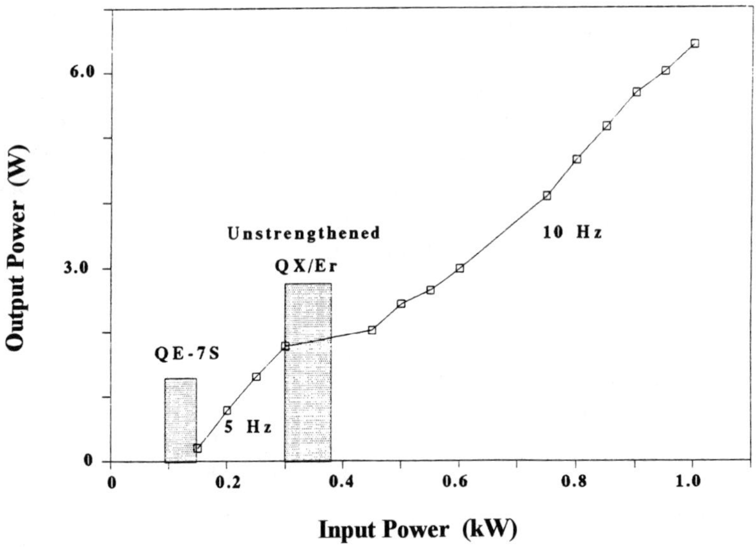

Figure 5 Laser Output vs. Pump Power for Strengthened QX/Er Glass

Table 4 lists the Q-switched results comparing QX\Er and QE-7S materials [3,9,10]. The longer pulse width at 15 Hertz is a result of thermal lensing, and can be easily compensated by modifications to the resonator.

Table 4. QX \Er and QE-7S Q-Switched Lasers

Parameter	QX\Er	QE-7S
Repetition Rate	15 Hz	3 Hz
Energy	5 mJ	5 mJ
Pulse Width	55 ns	30 ns
Q-Switch method	Spinning Prism	Spinning Prism

The performance levels represented here are considered a plateau in terms of further improvements in performance with these glasses. Any significant improvements beyond those represented will most likely have to be found in other glass formers. For improvements in efficiency, it appears that one would move to the telluride or chloride glasses. Figure 6 illustrates the relative positions of a number of glass formers, with respect to their cross section.

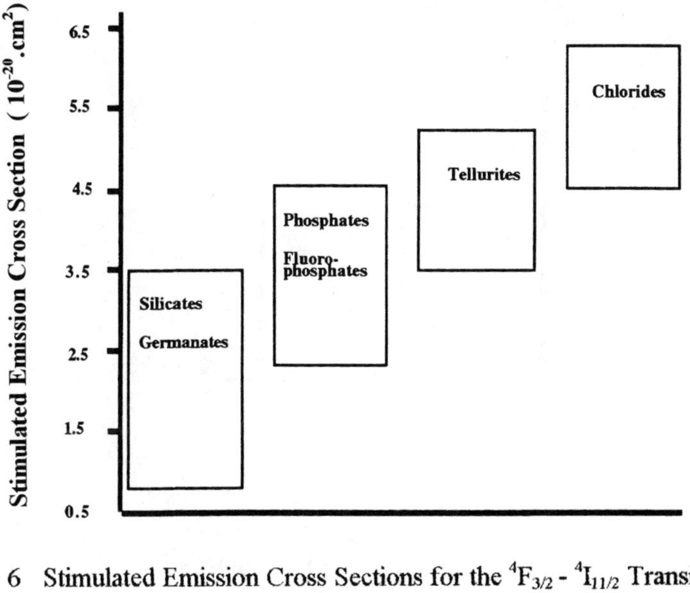

Figure 6 Stimulated Emission Cross Sections for the $^4F_{3/2}$ - $^4I_{11/2}$ Transition of Nd^{3+}

This information is gleaned primarily from the LLNL publication cited earlier [4].

2 Micron Glass Lasers

With the increasing interest in IR Lasers, Kigre developed a germanate laser glass to investigate laser glasses in the 2 to 3 micron region. The result was a glass with an index of refraction of 1.65 and a coefficient of thermal expansion of 65 x 10-7. Initially doping with holmium and thulium resulted in a laser glass with an output at 2 μm. Previous attempts at lasing Ho^{3+} in a silicate glass were reported in the 1960's at cryogenic temperatures. There is also a report of lasing of Tm^{3+} in a silicate glass back in 1967, again at very low temperatures. The performance of Ho^{3+} and Tm^{3+} in silicates is limited by the high multiphonon energy. Figure 7 shows the energy levels of Ho^{3+} and Tm^{3+}. Using this new germanate composition, the performance of Ho^{3+} was investigated at room temperature.

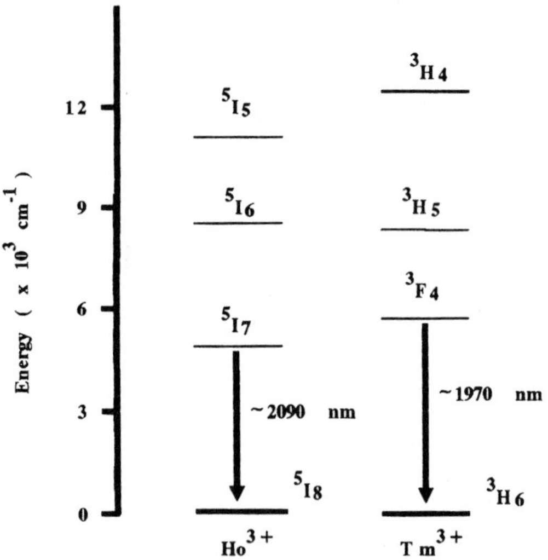

Figure 7 Energy Levels of Ho^{3+} and Tm^{3+}.

The results of that investigation; Ho^{3+} laser performance, lasing from the Ho^{3+} , $^5I_7$ - $^5I_8$ electronic transitions at 2.09 μm at room temperature are illustrated in Figure 8. Here the laser output energy at 0.2 Hz and 1 Hz is given as a function of the flashlamp input energy for Kigre's QG-108 glass rod [11]. This glass was sensitized by Er^{3+}, Yb^{3+} and Tm^{3+}.

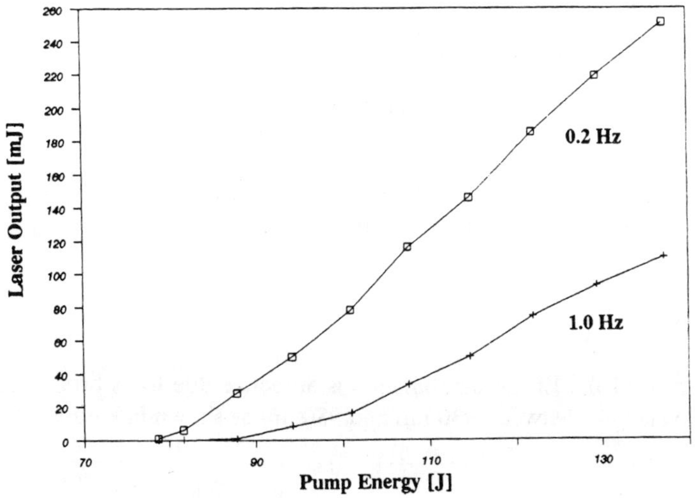

Figure 8 Output Energy as a Function of the Flashlamp Input Energy Using De-Ionized Water as the Coolant

Free running lasing tests have yielded a slope efficiency of 1.0%, a threshold of 51 Joules, and an output energy per pulse exceeding one Joule. Work is continuing on the Tm^{3+} analog. It was noted that the effective pump radiation was primarily in the IR. This was made obvious by the difference in laser performance observed when a highly absorbing coolant, water, is replaced by a coolant which absorbs less in the IR. Figure 9 demonstrates the performance at 0.2 Hz, of OG-108 glass with two different coolants; water and perfluoropolyether.

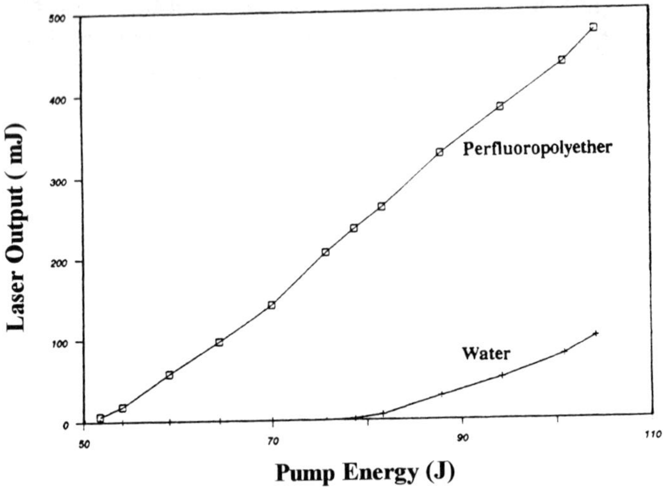

Figure 9 Comparison of Laser Characteristics for OG-108 Glass Using Different Coolants

Diode Pumping

Diode pumping of the Er-Yb combination is attractive due to its broad acceptance of pump wavelengths between 930 nm and 980 nm as shown in Figure 10.

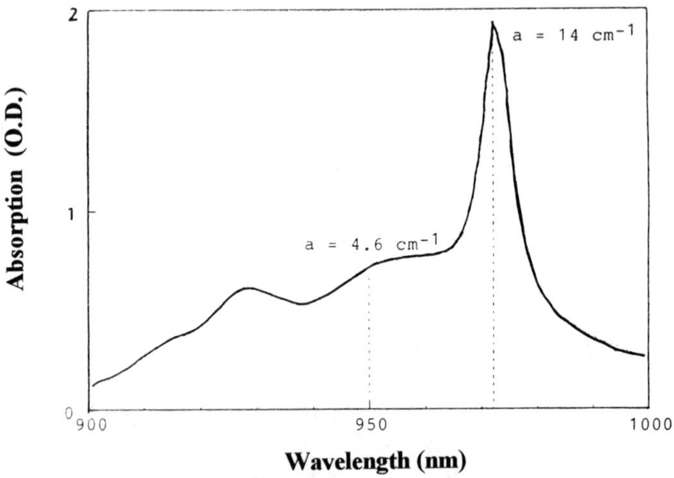

Figure 10 Absorption Spectrum of the QE-7 Yb,Er:Glass

The relevant well known energy levels are shown in Figure 11.

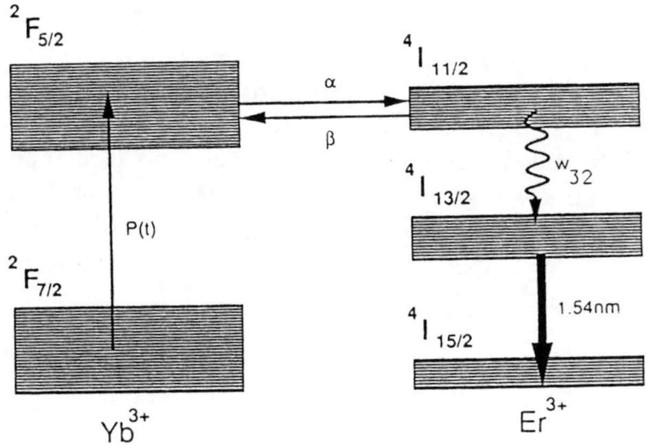

Figure 11 Yb,Er:Glass Energy Levels

Kigre, Inc., has supplied materials for diode pumping to several companies, laboratories and institutions including:

- The Instituto di Fisica del Politechnico, who reported a CW output of 30 mW in 1992.
- McDonnell Douglas, who reported 15 mJ Q-switched outputs at 1 Hz.
- The Night Vision and Electro-Optic Directorate, who reported a 20 mJ, 14% slope output when pumped at 970 nm,... and a Q-switched output of 0.3 mJ at 7 Hz.

Optimization of doping levels and pumping geometries continues. Efforts in our own labs are just beginning. We expect to see diode pumped laser glasses at Er^{3+}, Ho^{3+} and Tm^{3+} wavelengths in the near future.

Visible And Up-Conversion Laser Glasses

Current and future efforts, in the visible region, center around investigations of suitable glass hosts for Cu^{+} lasing at approximately 500 nm; Tb^{3+} lasing at approximately 550 nm; and Sm^{3+} lasing at approximately 600 nm. So far, we have had only limited success in this area and the search continues.

Up-conversion lasers using both Kigre's QX and Germanate hosts appear promising. Absorption and emission energy states of Er^{3+} and Tm^{3+} in Heavy Metal Fluoride (HMF) glasses exhibit several possibilities. In the case of the Er^{3+}, Hutchinson, at the U.S. Army Night Vision and Electro-Optics Directorate reported [12] visible upconversion fluorescence bands in diode-pumped QE-7 at 530 nm and 550 nm in the green and a band at 660 nm in the red, all of which were attributed to emission from higher excited states of the erbium ion. It is expected that the diode pumping work will uncover more promising up-conversion possibilities as work continues.

References

1. Handbook of Lasers, Ed. R J Pressley, CRC Press, Cleveland, pp363-364. 1971.
2. S Jiang, J D Myers, et al., Kigre Inc., "Chemically Strengthened Er^{3+}, Nd^{3+} Doped Phosphate Laser Glasses", SPIE Vol. 2376, 1995. Also Kigre Nd/QX data sheet 1995.
3. J D Myers, et al., Kigre.Inc., Athermal Laser Glass Composition with High Thermal Loading Capacity, US Patent No. 5,322,820, 1994. Also see J D Myers, et al., Kigre, inc., Athermal Laser Glass (Q-98 and Q-100), US Patent No. 4,333,848., 1981. Also see J D Myers, et al., Kigre, inc., Laser Phosphate Glass (Q-88), US Patent No. 4,075,120., 1978.
4. LLNL Publication M-95, Rev. 2, Vol. 1, Nov 1981.
5. V.C. Kravchenko and Yu. P. Rudnitskii, Sov. J. Quantum Electron, V9, pp 399-415, 1979.
6. John McMahon of Naval Research Laboratory, Private Communication, 1984.
7. Kigre, Inc., Lab Data on the patent first cited in ref. 3. 1994.
8. J D Myers, et al., Kigre, Inc., "High Repetition Rate Q-switched Erbium Glass Lasers", SPIE Vol. 1419 Eyesafe Lasers : Components, Systems and applications, 1991.
9. J D Myers, et al., Kigre, Inc., Sensitized Laser Glass (QE-7S), US Patent No. 4,770,811., 1988.

10. S Jiang and J D Myers, Kigre, Inc., D Rhonehouse, "Laser and Thermal Performance of a New Erbium Doped Phosphate Laser Glass", OE/LASE'94 Los Angeles Cal, Jan 1994.
11. S Jiang, J D Myers, R E Belford et al., Kigre.Inc., "Flashlamp Pumped Lasing of Ho:Germanium Oxide Glass at Room Temperature", Advanced Solid State Lasers, Tech. Dig., OSA, Washington DC, pp 329-331, 1994.
12. A Hutchinson, et al., U.S. Army Night Vision and Electro-Optics Directorate Private Communication 1995.

NEW SOLID STATE LASERS FROM THE ULTRAVIOLET TO THE MID-INFRARED

Stephen A. Payne, William F. Krupke, Raymond J. Beach, Laura D. DeLoach, Mark A. Emanuel, Christopher D. Marshall, Ralph H. Page, Kathleen I. Schaffers, Jay A. Skidmore
Lawrence Livermore National Laboratory
University of California
P.O. Box 5508
Livermore, CA 94551

Abstract

We discuss three new laser materials that offer improved access to the ultraviolet, near infrared and mid-infrared spectral regions. In order for each of these materials to have been identified, a particular hurdle needed to be overcome with respect to the fundamental laser physics impacting the material. In the case of the 280-320nm Ce:LiSAF laser, the main issue is the need to reduce the loss associated with excited state absorption, while for 1047nm Yb:S-FAP it is the ground state absorption at the laser wavelength that must be minimized. Cr:ZnSe has been down-selected from a number of potential candidates which could lase in the 2200-3000nm region, in order to mitigate the detrimental impact of nonradiative decay. In all three cases we discuss how appropriate consideration of fundamental concerns has led to the identification and understanding of the new laser system.

1. Introduction. One of the major goals facing the laser community has been to develop means of conveniently producing laser light from the mid-infrared to the ultraviolet. In the past there has been a very broad range of wavelengths demonstrated, but most of these laser sources have not proved to be practical devices. It turns out, however, that the insights derived from solid state physics, material science, and laser physics can now be collectively engaged to develop new laser media that are of practical interest to the user community. In other words, while the new laser media discussed in this paper are at a pre-commercial stage of development, we believe that these particular material systems offer exceptional prospects for future deployment in real laser applications. Now, although the overall utility of a laser material is potentially impacted by a very large number of factors other than the fundamental physics (e.g. crystal growth, thermomechanical properties, cost, etc.), this paper will address only the fundamental areas at this time (references quoted herein can be reviewed for greater detail).

Figure 1 reveals the fundamental physics that can affect the utility of a laser material. While there are many other issues of crucial importance to the gain medium, this article will focus on three aspects of the laser physics: excited state absorption, ground state absorption at the laser wavelength, and nonradiative decay of the metastable upper laser level, as illustrated in the figure.

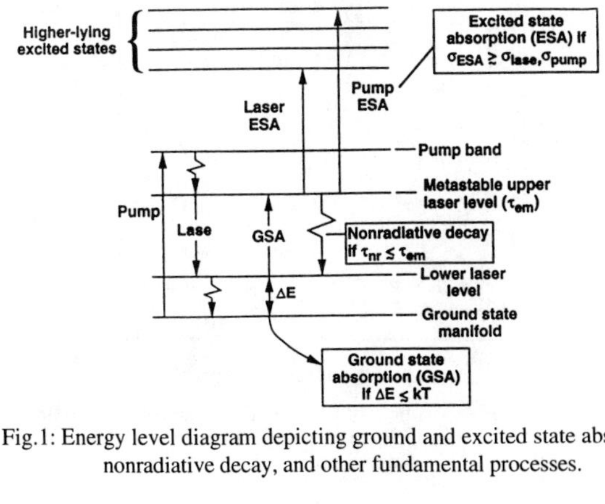

Fig.1: Energy level diagram depicting ground and excited state absorption, nonradiative decay, and other fundamental processes.

2. Ce:LiSAF lasers.[1] Excited state absorption (ESA) can detract from both the gain and the pumping efficiency, depending on whether the ESA exists at the laser or the pump wavelength. In the case of Ce:LiSAF (Ce^{3+}-doped into $LiSrAlF_6$), it turns out that both mechanisms are important to the performance and efficiency of the laser material. This situation can be addressed by first examining the absorption and emission spectra contained in Fig. 2. The absorption spectra may be interpreted to indicate that the material can be pumped in the range of 255-275nm, and that it should be relatively insensitive to the polarization of the pump source. (The $LiSrAlF_6$ host is uniaxial in character, and therefore exhibits the so-called p unique axis and the s plane perpendicular to it.) Furthermore, the emission spectra offer similar cross sections in the two polarizations, suggesting that we should expect similar gain in both cases.

As revealed in Fig. 3 however, it is apparent that the selection of either p or s polarized pump or probe light has a dramatic impact on the performance of the laser. The experiment employed to obtain the data of Fig. 3 is that where separate pump (266nm Nd:YAG) and probe (290nm dye) laser beams are arranged to impinge on the sample, and where the probe field is weak compared to that of the pump. This plot of laser gain versus the pump fluence reveals that the most favorable situation is one where the pump and probe are both π-polarized. For the cases where either the pump or the probe is σ-polarized we observe a very substantial decrease in the amount of gain that is achieved. These observations can be explained by suggesting that the loss process involves ESA at the pump and the probe wavelengths, and that the ESA is strongly σ-polarized. In this way, it is advantageous for the pump source and the laser to be π-polarized, to avoid the ESA losses. We conclude that it is feasible to overcome most of the ESA problem in the case of Ce:LiSAF by carefully choosing the polarizations of the laser beams.[1]

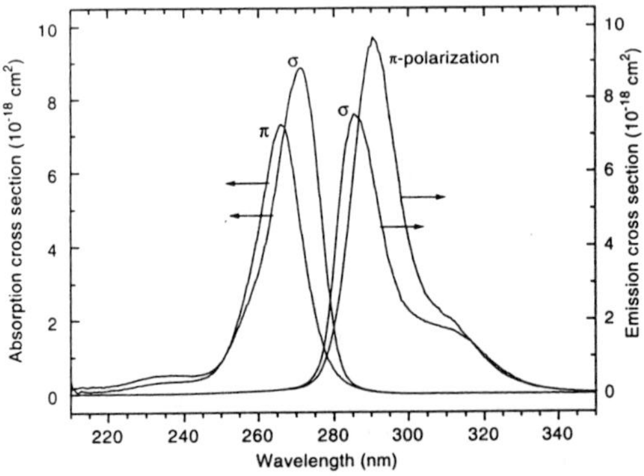

Fig.2: Absorption and emission spectra of Ce:LiSAF

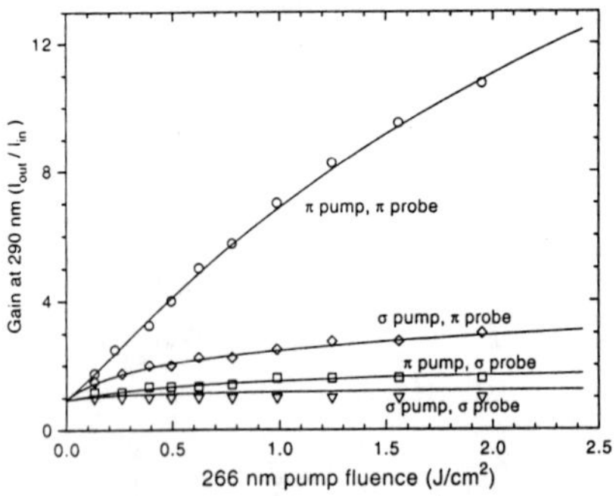

Fig.3: Pump-probe experiment of Ce:LiSAF, illustrating importance of the polarization in determining the level of gain that is present.

It is possible to understand the nature of the broad ESA feature that is under the absorption and emission bands. On the basis of previous studies of rare earth ions, the ESA has become understood to arise from a transition in which an electron is transferred from the Ce^{3+} ion to the conduction band of the $LiSrAlF_6$ host.[2] Upon examination of the host crystal it is noticed that the Sr^{2+} cations are arranged in planes perpendicular to the c axis of the lattice. Since the lowest conduction band component is most likely to be composed of the Sr ions, it may be suggested that the ESA is expected to be σ-polarized (or along the direction of the electron transfer step).

3. Yb:S-FAP lasers.[3-7] We have pursued the identification of a new class of Yb-lasers. Since Yb generally lases at nearly the same wavelength as the much better characterized Nd-lasers, one would naturally inquire as to what the potential advantages of ytterbium are over the neodymium-based alternatives. The main benefits are derived from the longer emission lifetime (3-5x) and the reduced quantum defect (0.3-0.5x). Since the emission lifetime impacts the energy storage that is possible for the material, ytterbium-doped media are roughly capable of storing 3 to 5x more energy for a given pump rate. For situations where the delivered energy per pulse from the laser is a specified value, and where the cost of the pump is important (e.g. as for laser diodes), the Yb-doped media can present the greatest advantage over Nd. The quantum defect between the pump and laser photons is also noteworthy, since the smaller difference characteristic of Yb leads to reduced heating of the gain medium. On the other hand, lasers based on the ytterbium ion tend to operate as quasi-three-level systems so the level of ground state absorption at the laser wavelength is an important aspect of the laser physics. Furthermore, Yb-doped crystals and glasses tend to exhibit rather low cross sections, rendering it difficult to efficiently extract the stored energy from amplifiers (or alternatively, reach threshold in laser oscillators). We have sought to identify new laser materials that would emphasize the favorable characteristics of Yb while minimizing the potential problems.

We have studied a variety of Yb-doped crystals in order to identify those materials that appear most promising with respect to the ground state absorption and the gain cross section. In this vein the figure-of-merit plot pictured below in Fig. 4 was devised. Here the parameter indicated as I_{min} serves the role of quantifying the difficulty of overcoming the ground state absorption at the laser wavelength. (It is actually the product of the pump saturation intensity and the fraction of ions that must be inverted, to yield transparency at the laser wavelength.) A simple explanation of I_{min} is that it is the minimum absorbed pump intensity required to reach threshold in an otherwise lossless resonator (using a lightly doped sample). The objective of our search as defined by Fig. 4 is to identify the Yb-doped materials that possess the maximal gain cross section and the smallest value of I_{min}. Upon inspection of the crystals characterized and included in the figure, we see that the family having the apatite structure is the most promising laser material candidates, including $Ca_5(PO_4)_3F$, $Sr_5(PO_4)_3F$, mixtures of these two compounds, and the related vanadate derivatives.

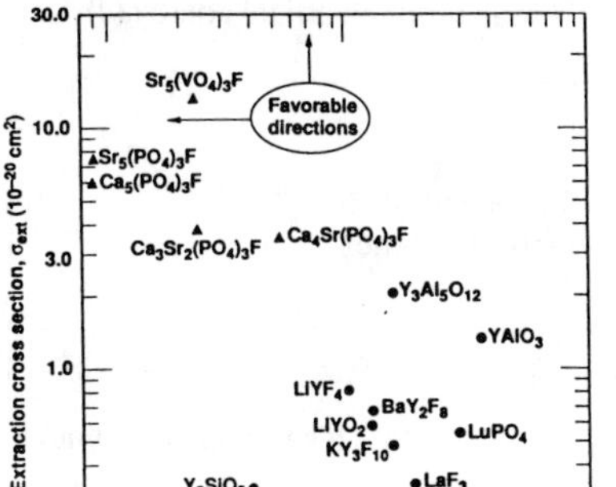

Fig.4: Figure-of-merit plot of the gain cross section against the minimum pump intensity.

The absorption and emission spectra of Yb in strontium fluorophosphate (Yb:S-FAP) are drawn in Fig.5. Here the pump line is seen to be at 900nm, while the laser line is at 1047nm. It's interesting to see that the absorption level at 1047nm is less than 5% of the gain cross section; this useful situation is reflected by the low value of I_{min} seen in the figure-of-merit plot of Fig.4. The spectra of Fig. 5 are striking in terms of the sharpness of the pump and laser lines, in that they appear to be subject to very little phonon broadening compared to nearly all other Yb spectra in crystalline media.

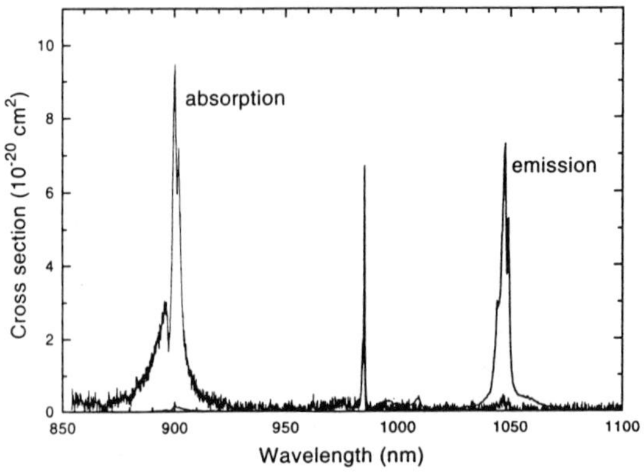

Fig.5: Absorption and emission spectra of Yb:S-FAP.

We have explored the laser performance of Yb:S-FAP and its compositional analogs using laser pumping (Ti:sapphire) and have measured efficiencies as high as 70%. Having scientifically confirmed the utility of the Yb:apatite lasers, we next developed a laser diode array pumped Yb:S-FAP oscillator and operated the 3x3x30mm rectangular element at over 13 Watts of output power, as shown below in Fig. 6. To construct this laser we fabricated a diode array that produces 3 kWatts of peak power, and then attached microlenses to the laser diode facets to collimate the light in the fast axis. A lens duct was employed to concentrate the light and direct it into the laser rod so that it was captured (by total-internal reflection). The laser rod was held between two water-cooled copper blocks to manage the waste heat that was generated. The overall electrical efficiency of the system is more than 10%. We believe that it should be possible to attain about twice this power and efficiency by securing higher quality Yb:S-FAP laser rods.

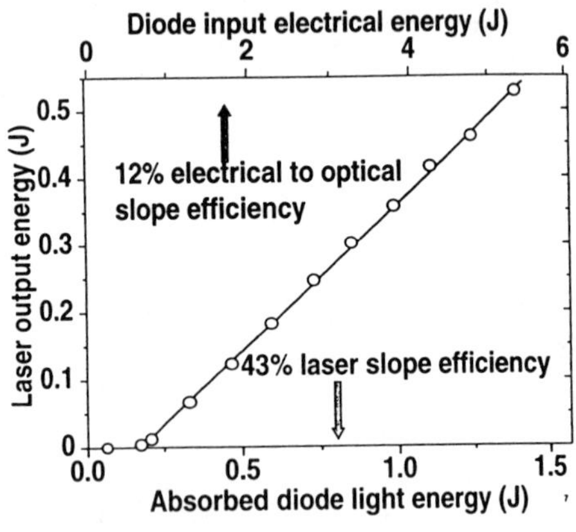

Fig.6: Laser performance of a laser diode array pumped Yb:S-FAP oscillator.

4. Cr:ZnSe lasers.[8] One of the challenging problems in laser science has been the development of systems that can generate tunable laser light in the mid-infrared spectral region. In order to accomplish this objective, optical parametric oscillators (OPOs) have been pursued. OPOs operate on the basis of nonlinear interactions, and therefore exhibit substantial sensitivity in their operating characteristics. Although these systems have proved to be useful and versatile, some technical circumstances call for the availability of direct tunable mid-infrared radiation from a diode-pumped solid state laser, since the system is expected to be more robust. Unfortunately no efficient lasers existed to permit direct generation of tunable infrared laser radiation. The main impediment to identifying mid-infrared tunable laser materials had been the predominating influence of nonradiative decay in most optical materials, which leads to a low emission quantum yield and high threshold of laser operation. We have uncovered a new class of laser materials that have a high quantum yield in the mid-infrared, based on Cr^{2+}-doped ZnSe crystals (and other II-VI hosts).

The absorption and emission spectra of Cr:ZnSe are displayed in Fig. 7, where it is seen that the features are quite broad. We can pump the Cr^{2+} ion at about 1800nm, and tunable radiation should be available from 2200-3000nm. The broadness of the spectra is a consequence of the strong interactions of the nearest-neighbor selenide anions with the 3d-3d transitions of chromium. Since the concept of employing divalent chromium as the laser ion as well as using ZnSe as the host medium were both previously untested, we sought to perform a scientific demonstration exercise where it was conclusively shown that Cr:ZnSe could offer gain and reach threshold in a laser cavity. To accomplish this, we used the "Cobra" laser from Schwartz Electro-Optics to pump a crystal in a 20cm confocal oscillator, and achieved laser action as pictured below in Fig.8. The untuned output from the laser was at 2350nm, as would be expected from the emission peak position shown in Fig.7.

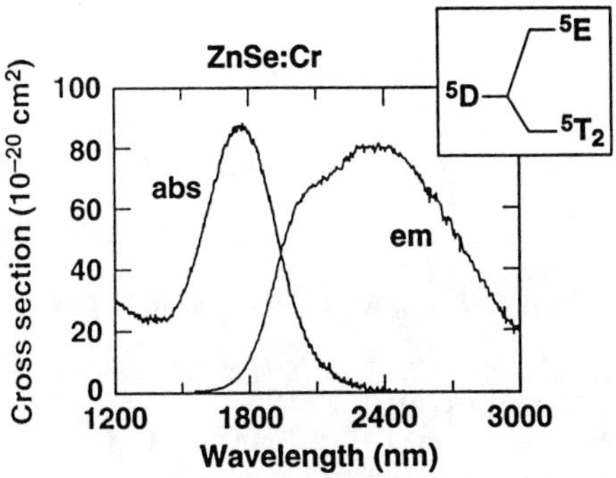

Fig.7: Absorption and emission spectra of Cr:ZnSe, revealing the pump band at 1800nm and the emission band peak at 2350nm.

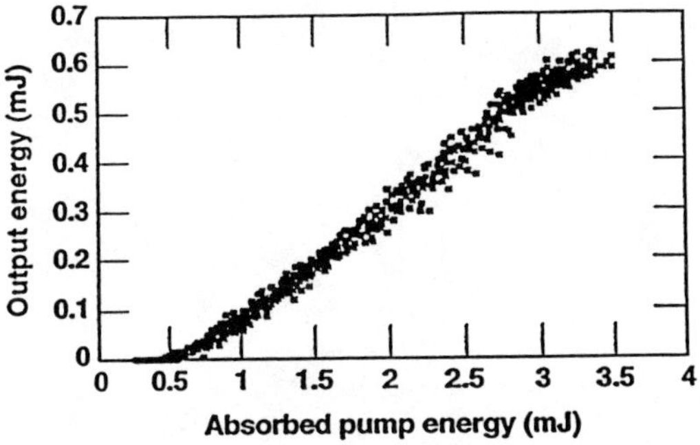

Fig. 8: Scientific demonstration of laser action for Cr:ZnSe pumped at 1700nm and generating laser light at 2350nm.

From the data displayed in Figs. 7 and 8, we can conclude that a new class of lasers has been identified, which is exceptionally promising for several reasons. First, this laser material and its chemical analogs are the first systems found to emit efficiently in the mid-infrared with a broad spectral band. Second, the ZnSe host is thermomechanically robust and should serve adequately in many types of laser systems. Third, a substantial commercial element already exists regarding the manufacture of ZnSe optics. Finally, the preferred pump wavelength of about 1800nm it accessible on the basis of known laser diode structures. We believe that the Cr:ZnSe laser demonstration is the beginning of a new approach to generating mid-infrared laser light.

5. Summary. We have discussed three novel laser materials first lased at Lawrence Livermore National Laboratory which offer access to the ultraviolet, infrared, and mid-infrared spectral ranges. The laser physics of each of these materials was addressed at a fundamental level in order to provide optimal performance. For the case of the 280-320nm tunable Ce:LiSAF laser, excited state absorption at both the pump and laser wavelengths was determined to be important, although we found that it was possible to mitigate its impact by employing π-polarized light. Yb:S-FAP operates at 1047nm (which is similar to the well-known Nd-lasers), but provides a substantially longer storage time. The special advantage of Yb:S-FAP among potential Yb-lasers is related to its low level of ground state absorption loss at the laser wavelength. The Cr:ZnSe laser represents a new approach to directly generating tunable mid-infrared (2200-3000nm) radiation, on the basis of a material that emits with a high quantum yield.

This work was performed under the auspices of the U.S. Department of Energy by Lawrence Livermore National Laboratory under Contract No. W-7405-Eng-48.

1. C. D. Marshall, J. A. Speth, S. A. Payne, W. F. Krupke, G. J. Quarles, V. Castillo, and B. H. T. Chai, J. Opt. Soc. Am. B 11, 2054-2065 (1994).
2. J. K. Lawson and S. A. Payne, Phys. Rev. B 47, 14003-14010 (1993).
3. L. D. DeLoach, S. A. Payne, L. L. Chase, L. K. Smith, W. L. Kway, and W. F. Krupke, IEEE J. Quantum Electron. 29, 1179-1191 (1993).
4. S. A. Payne, L. K. Smith, L. D. DeLoach, W. L. Kway, J. B. Tassano, and W. F. Krupke, IEEE J. Quantum Electron. 30, 170-179 (1994).
5. L. D. DeLoach, S. A. Payne, L. K. Smith, W. L. Kway, and W. F. Krupke, J. Opt. Soc. Am. 11, 269-276 (1994).
6. S. A. Payne, L. D. DeLoach, L. K. Smith, W. L. Kway, J. B. Tassano, W. F. Krupke, B. H. T. Chai, and G. Loutts, J. Appl. Phys. 76, 497-503 (1994).
7. C. D. Marshall, S. A. Payne, L. K. Smith, H. T. Powell, W. F. Krupke, and B. H. T. Chai, IEEE J. Selected Topics Quantum Electronics 1, 67-77 (1995).
8. L. D. DeLoach, R. H. Page, G. D. Wilke, S. A. Payne, and W. F. Krupke, submitted to J. Quantum Electron.

95-085

INFRARED TO VISIBLE UPCONVERSION IN Er^{3+} DOPED LEAD-GERMANATE GLASSES

Z. Pan, K. Dyer, A. Loper, and S. H. Morgan
Physics Department, Fisk University
Nashville, TN 37208

ABSTRACT

The upconversion mechanism of Er^{3+} ions has been studied for $xPbO(89\text{-}x)GeO_2 10CaO 1Er_2O_3$ (x = 15, 27, 50) glasses. Intense green emission was observed at room temperature due to $^4S_{3/2} \rightarrow {}^4I_{15/2}$ transition excited by a cw near-infrared laser beam at 797 nm. This green emission showed a similar intensity for samples with different lead concentrations. A red emission $^4F_{9/2} \rightarrow {}^4I_{15/2}$ increases with decreasing lead concentration. The bright green emission is attributed to the excited level absorption while the red emission is partially attributed to the energy transfer between excited ions owing to their Coulomb interaction.

INTRODUCTION

In this article, we report the infrared to visible upconversion observed in Er^{3+} doped $PbO\text{-}GeO_2$-CaO glass system (Er^{3+}:LGC), excited by a near-infrared laser beam at 797 nm. The glasses were stable by the conventional casting procedure, and no devitrification was observed. As previously reported, by further optimizing the composition, lead-germanate based glasses may be suitable for developing low loss, high strength, and low cost fiber devices.[1] The maximum vibrational energy of lead-germanate glass is about 820 cm^{-1}, intermediate between those of silicate (~1150 cm^{-1}) and fluoride (~580 cm^{-1}) based glasses. Previously we have reported the infrared to visible upconversion in $27PbO{\cdot}63GeO_2{\cdot}10CaO{\cdot}$ glass with different Er^{3+} concentration.[3] In this article we report the influence of the different PbO composition.

EXPERIMENTAL

The glasses were prepared by mixing PbO, $CaCO_3$, GeO_2, and Er_2O_3 powders in 30 g batches. PbO and $CaCO_3$ were reagent grade while GeO_2 was of 99.9999% purity (Eagle Picher Research Lab) and Er_2O_3 was of 99.99% purity (Alfa, Johnson Matthey). Three glasses in the form $(89\text{-}x)GeO_2\, xPbO\, 10CaCO_3\, 1Er_2O_3$ were fabricated with the x = 15, 27, 50, respectively. The mixtures were melted in fused silica crucibles at a temperature of 1050 C and held for 25 minutes. The melts were cast onto a copper plate and pressed by another copper plate from the top, forming a glass plate of about 1.0 mm in thickness. The glasses were subsequently annealed at 450 C for 50 minutes and allowed to cool to room temperature in the furnace. All three glasses were stable and no devitrification was observed.

The upconversion fluorescence spectra were measured at room temperature using a Spex Model 1403 double grating spectrometer with a cooled RCA3134 photomultiplier and photon-counting system. The excitation source used was a Lexel cw near-infrared Ti:sapphire laser tuned at 797 nm. The energy levels shown in Fig. 1 were obtained from the absorption spectra.

RESULTS AND DISCUSSION

The room temperature upconversion luminescence spectra, excited at 797 nm, and the luminescence spectra, excited at 488 nm, are shown in Fig. 2 for an Er^{3+}:LGC glass with 15 mol% PbO and an Er^{3+}:LGC glass with 50 mol% PbO. It is found that the dominant emission is in the green corresponding to $^4S_{3/2} \rightarrow {^4I_{15/2}}$ transition of Er^{3+}. The green emission in the region 514 nm to 573 nm accounts for more than 80 % of the total emitted light in the spectral region studied (28,000 to 11,300 cm^{-1}). The bright green fluorescence is visible to the naked eye even for 20 mw of the pump laser power at room temperature.

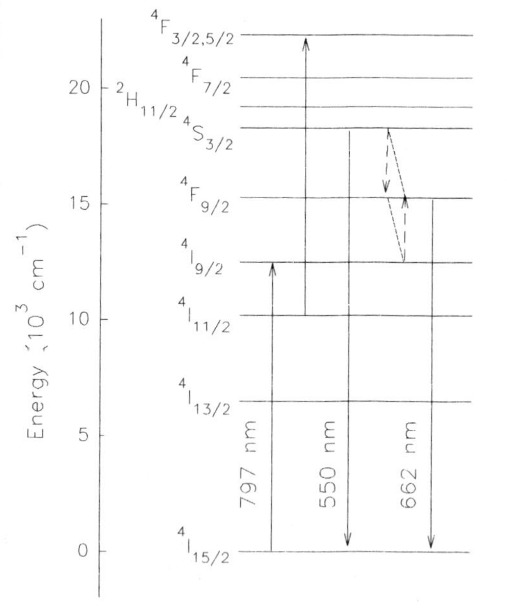

Figure 1: Energy-level diagram and a possible mechanism of infrared to visible upconversion for Er^{3+} doped lead-germanate glasses under 797 nm excitation.

The upconversion spectra also show a red emission band at 662 nm, a near-infrared band at 855 nm, and a weak blue emission at 410 nm. According to the energy diagram shown in Fig. 1, these three emission bands are identified as the $^4F_{9/2} \rightarrow {^4I_{15/2}}$, $^4S_{3/2} \rightarrow {^4I_{13/2}}$, and $^2H_{9/2} \rightarrow {^4I_{15/2}}$ transitions, respectively. Our results show that (1)the intensity branching ratio of the upconversion red emission to green emission is larger than that by direct excitation; (2) this branching ratio decreases with increasing PbO concentration.

There are two basic upconversion mechanisms: excited level absorption and energy transfer between excited ions.[2] As discussed in our previous paper, the green emission is basically a two-step upconversion process. The two sequential excitations are $^4S_{15/2} \rightarrow {}^4I_{9/2}$ and $^4S_{11/2} \rightarrow {}^4F_{3/2,5/2}$ transitions.[3] The ions in the $^4F_{3/2,5/2}$ levels undergo multiphonon relaxation through $^4F_{7/2}$ to $^2H_{11/2}$ and $^4S_{3/2}$ levels. Because the energy gap below the $^4S_{3/2}$ level is larger, the bright green fluorescence was emitted through $^4S_{3/2} + {}^2H_{11/2} \rightarrow {}^4I_{15/2}$ transitions.

The red emission is identified as the $^4F_{9/2} \rightarrow {}^4I_{15/2}$ transition. The $^4F_{9/2}$ level can be populated through two paths: (1) from $^4S_{3/2}$ level through multiphonon relaxation; (2) via the energy transfer between excited ions owing to their Coulomb interaction. The energy transfer due to the Coulomb interaction occurs for Er^{3+} ion concentration $\geq$ 0.5 at. %. The concentration dependence has been discussed in our previous paper.[3] The two coupled transitions are identified as $^4S_{3/2} \rightarrow {}^4F_{9/2}$ and $^4F_{9/2} \leftarrow {}^4I_{9/2}$. The energy gaps between the two coupled transitions are matched well with a small red shift, which is favorable to this kind of transfer.[4] The transfer rate is proportional to the product of populations at $^4S_{3/2}$ level and $^4I_{9/2}$ level. Because the population at $^4I_{9/2}$ level is much higher by 797 nm excitation than by 488 nm excitation, a higher energy transfer rate is expected by the 797 nm excitation. We observed a larger red to green intensity ratio in the upconversion spectrum compared to that by the direct excitation. This result indicates that the energy transfer process is important in the upconversion red emission, consistent with the results of a previous study for different Er^{3+} concentration.[3]

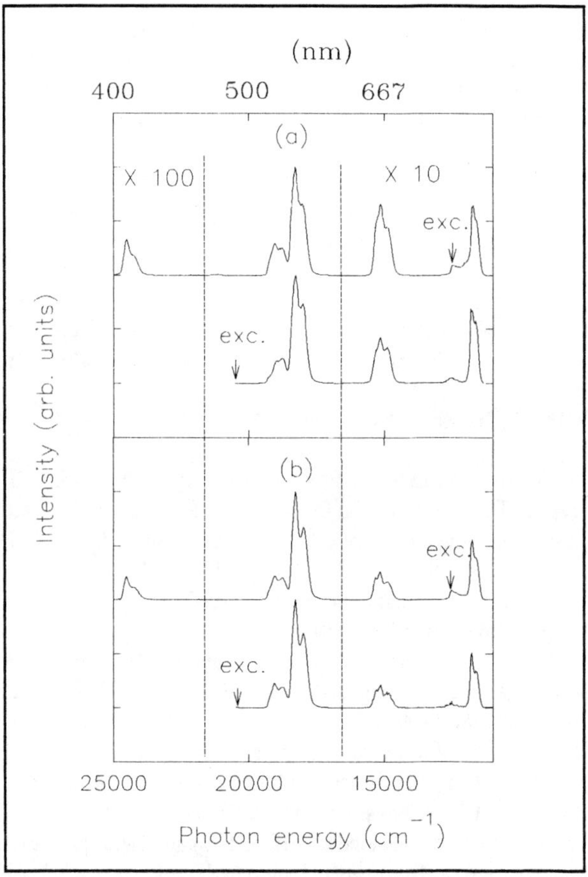

Figure 2: Upconversion luminescence excited at 797 nm and luminescence excited at 488 nm, (a) 15 mol% PbO glass, (b) 50 mol% PbO glass.

The decrease in the red to

green branching ratio by increasing PbO concentration can be attributed to the effect of the lead on the GeO_2 glass structure. The PbO distorts the tetrahedral GeO_4 network, and increases the number of nonbridging oxygens in the glass network. The lead cations in the glass may have two kinds of states, a covalent one and an ionic one, dependent on the lead concentration.[5] The Er^{3+} ions in the glass are surrounded by nonbridging oxygens of GeO_4 tetrahedra and lead-oxygen polyhedra. The symmetry of rare-earth ligand field may be modified by increasing PbO concentration. As suggested by a previous study that the branching ratio of upconversion emissions is related to the symmetry of rare-earth ligand field.[6]

CONCLUSION

We have reported the room temperature infrared to visible upconversion in Er^{3+} doped lead-germanate glass with different PbO concentrations, excited by 797 nm laser beam. The dominant emission is $^4S_{3/2} \rightarrow {}^4I_{15/2}$ in the green by a two-step upconversion process. A red emission $^4F_{9/2} \rightarrow {}^4I_{15/2}$ is partially due to the energy transfer among excited Er^{3+} ions.

The branching ratio of red to green upconversion luminescence decreased by increasing PbO concentration in glass. Control of the upconversion luminescence branching ratio may be possible by adjusting the PbO concentration in the host.

We acknowledge the financial support of NASA through grant NAGW-2925 and the Tennessee Space Grant Consortium, and NSF under REU grant DMR-9322437.

REFERENCES

[1] J. Wang, J. R. Lincoln, W. S. Brocklesby, R. S. Deol, C. J. Mackechnie, A. Pearson, A. C. Tropper, D. C. Hanna, and D. N. Payne, "Fabrication and optical properties of lead-germanate glasses and a new class of optical fibers doped with Tm^{3+}", J. Appl. Phys. **73**, 8066 (1993).
[2] J. P. van der Ziel, L. G. Van Uitert, W. H. Grodkiewicz, and R. M. Mikulyak, "1.5 μm infrared excitation of visible luminescence in $Y_{1-x}Er_xF_3$ and $Y_{1-x-y}Er_xTm_yF_3$ via resonant-energy transfer", J. Appl. Phys. **60**, 4262 (1986).
[3] Z. Pan, S. H. Morgan, A. Loper, V. King, B. H. Long, and W. E. Collins, "Infrared to visible upconversion in Er^{3+}-doped-lead-germanate glass: Effects of Er^{3+} ion concentration", J. Appl. Phys. **77**, 4688 (1995).
[4] K. Huang, "Lattice relaxation and multiphonon transitions", Contemp. Phys. **22**, 599 (1981).
[5] Z. Pan, D. O. Henderson, and S. H. Morgan, "A Raman investigation of lead haloborate glasses", J. Chem. Phys. **101**, 1767(1994).
[6] S. Tanabe and T. Hanada, "Effect of ligand field on branching ratio of ultraviolet and blue upconversion of Tm^{3+} ions in halide and oxide glasses", J. Appl. Phys. **76**, 3730(1994).

THIRD-ORDER OPTICAL NONLINEARITY OF TeO_2-PbO-GeO_2 GLASSES

You Song Kim, Weon Hyo Kim and Jong Heo

Department of Materials Science and Engineering
Pohang University of Science and Technology, Pohang 790-784, Korea

ABSTRACT

Nonlinear optical properties of TeO_2-PbO-GeO_2 system were investigated and their correlation with the structural modification was discussed. Third-order optical nonlinear susceptibility, $\chi^{(3)}$, ranged between 10.7×10^{-13} esu and 4.95×10^{-13} esu, and decreased with increasing GeO_2 addition. Analyses of the Raman spectra suggested that these glasses are mainly composed of [TeO_4], [TeO_3] and [GeO_4] structural units. It was concluded that polarizability of Pb^{2+} ion in TeO_2-PbO glasses overwhelmed the effect of structural modification of [TeO_4]tbp→[TeO_3]tp. In TeO_2-PbO-GeO_2 glasses addition of GeO_2 resulted in the decrease of $\chi^{(3)}$ values. This behavior was attributed to the formation of [GeO_4] polyhedra at the expense of [TeO_n] polyhedra and Pb^{2+} ions which normally showed a higher contribution to $\chi^{(3)}$ than [GeO_4] polyhedra.

INTRODUCTION

Optical glasses are potential candidate materials for all-optical switching devices based on the third-order optical nonlinearity. Glasses generally belong to the category of having low optical nonlinearity, but provide advantages in terms of ease of fabrication, high transparency and fast switching time which can easily offset their drawbacks[1]. TeO_2-based glasses, among many oxide glasses investigated, seem to be most promising with their third-order optical nonlinear susceptibilities in the range of 10^{-13}-10^{-12} esu[2]. It has been suggested that the unusually large optical nonlinearity of TeO_2-based glasses originates from the presence of lone-pair electrons and their unique molecular structure.

In the present work, (TeO_2-PbO)-based glass system was selected for large optical nonlinearity. GeO_2 was added as a third component in order to improve glass-forming ability of the binary glasses. Raman spectra were examined for both binary and ternary glasses in order to investigate the relationships between $\chi^{(3)}$ and molecular structures.

EXPERIMENTAL PROCEDURE

Glasses with compositions of (100-x)(80TeO_2-20PbO)-xGeO_2 (x=0, 5, 10, 15, 20 mol%) and (100-x)TeO_2-xPbO (x=5, 10, 15, 20, 25 mol%) were prepared by the conventional melting and quenching method. Refractive indices of glasses were measured by an ellipsometer at the wavelength of 633nm. Third-order optical nonlinear susceptibilities, $\chi^{(3)}$, were determined by third harmonic generation(THG) method as described elsewhere[3]. The laser beam of 1.9μm wavelength with 10ns duration was used as a pumping source. Raman spectra were obtained with double spectrometer using the unpolarized 514.5nm line of an Ar^+ laser(100mW) at room temperature. Measured Raman spectra were reduced with following relation, $I_R = I\omega[n(\omega)+1]^{-1}$ where I_R and I are the reduced and observed Raman scattering intensities at wavenumber ω, respectively. $n(\omega)=[\exp(\hbar\omega/kT)-1]^{-1}$ is Bose-Einstein thermal population factor.

RESULTS AND DISCUSSION

Third-order optical nonlinear susceptibilities, $\chi_{1111}^{(3)}(-3\omega, \omega, \omega, \omega)$, were calculated by the following equation[4];

$$\chi^{(3)} = \chi_s^{(3)}\left(\frac{l_{c,s}}{l_c}\right)\left(\frac{I_{3\omega}}{I_{3\omega,s}}\right)^{1/2}\left(\frac{n+1}{n_s+1}\right)^4 \tag{1}$$

where l_c, $I_{3\omega}$ and n are the coherent length, third harmonic intensity and refractive index of the sample, respectively. Subscript s represents a reference sample. Silica glass with $\chi_s^{(3)}=2.8\times10^{-14}$ esu, $n_s=1.458$ and $l_{c,s}=18.1\mu m$ was used as a reference sample for the present work. $\chi^{(3)}$ values obtained from Eq. (1) and the refractive indices were summarized in Table I.

Table I. Optical Properties of (100-x)(80TeO_2-20PbO)-xGeO_2 Glasses

x(mol%)	n_{633}	$\chi^{(3)}$ ($\times10^{-13}$esu)
0	2.18	10.7
5	2.12	9.68
10	2.10	8.84
15	2.08	7.23
20	2.04	4.95

$\chi^{(3)}$ values ranged in 10.7-4.95$\times10^{-13}$ esu which are approximately 40-20 times of that of silica glass. Also, the determined $\chi^{(3)}$ values of glasses decreased with increasing GeO_2 content. It may be attributed to the decrease in the polarizability of the chemical bonds forming the glasses which will be discussed in further detail later.

Reduced Raman spectra of TeO_2-PbO glasses are shown in Fig. 1 (a). There are three well-developed peaks at 720cm^{-1}, 660cm^{-1} and 465cm^{-1}. First two peaks were assigned to the $Te-O^-$ and Te=O stretching vibration of [TeO_3] trigonal pyramid(tp) and Te-O stretching vibration of [TeO_4] trigonal bipyramid(tbp), respectively[9]. The last peak at 465cm^{-1} is due to the either stretching or bending vibration of Te-O-Te linkages which were formed by vertex-sharing [TeO_4]tbp's and [TeO_3]tp's[5]. With an increasing PbO concentration, peaks at 660cm^{-1} and 465cm^{-1} became weaker, while one located at 720cm^{-1} became more intense. It indicates that [TeO_4]tbp were transformed to [TeO_3]tp with the addition of PbO. At the same time, intensity of the 465cm^{-1} peak decreased sharply which further suggested that the glass network became open structure with Pb^{2+} ions probably acting as network modifiers.

Fig. 1 (b) shows the reduced Raman spectra of the ternary TeO_2-PbO-GeO_2 glasses. Shape of the Raman spectra were similar to that of 80TeO_2-20PbO glass. Structure of GeO_2 glass usually consist of a continuous network of [GeO_4] tetrahedral units. When typical modifier oxides such as alkali oxides were added into GeO_2 glass, transformation of structural units from [GeO_4] tetrahedra to [GeO_6] octahedra occurred. For instance, Raman spectra for alkali germanate glasses containing 15 mol% of sodium oxide showed a well-resolved band related to [GeO_6] octahedra in the range of $600cm^{-1}$-$650cm^{-1}$[6]. It has been also speculated that such structural changes also occurred in lead germanate glasses. However, the maximum amount of [GeO_6] units formed in this system was approximately half of the amount in binary alkali germanate glasses with similar composition[7]. In the Raman spectra of TeO_2-PbO-GeO_2 glasses, however, no significant changes in the relative intensities of two peaks at $660cm^{-1}$ and $720cm^{-1}$ were observed. Therefore, it is considered that no [TeO_4]tbp↔[TeO_3]tp transition has occurred with GeO_2 addition, and [GeO_4] tetrahedra were main structural units formed in the glass network.

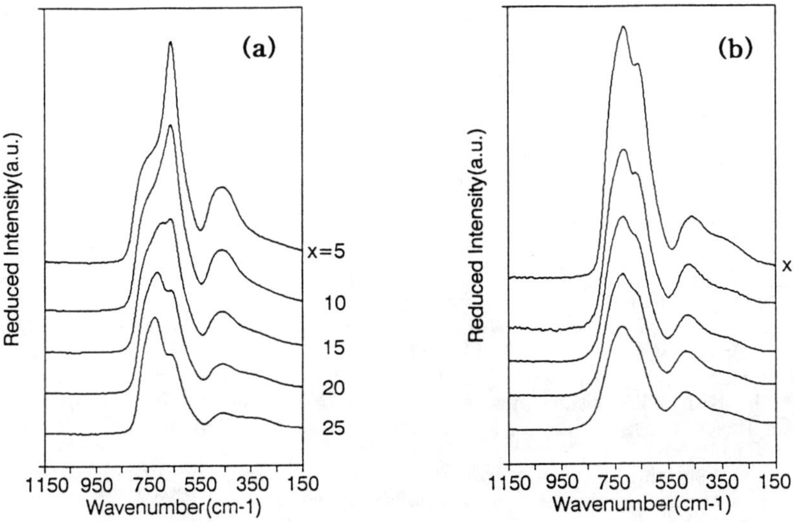

Fig. 1. Reduced Raman spectra of (a)(100-x)TeO_2-xPbO glasses, (b) (100-x)(80TeO_2-20PbO)-xGeO_2 glasses

In general, changes in the coordination state affect the bond energies and hence influence nonlinear polarizabilities of the chemical bonds. Yamamoto et al.[8] examined the structure-optical nonlinearity relationships for TeO_2-Li_2O glasses and suggested that the contribution of [TeO_3]tp to $\chi^{(3)}$ is smaller than that of [TeO_4]tbp. Kim et al.[9], on the other hand, reported $\chi^{(3)}$ values of TeO_2-PbO glasses increased from 14.1×10^{-13} esu to 16.4×10^{-13} esu with the addition of up to 20 mol% of PbO. As mentioned previously, addition of PbO to TeO_2 glasses induced the transformation of [TeO_4]tbp to [TeO_3]tp which would result in the negative effect on the optical nonlinearity. Increase in the determined $\chi^{(3)}$ values with PbO addition, therefore suggests that the contribution of Pb^{2+} ion to $\chi^{(3)}$ is larger than that of [TeO_4]→[TeO_3] transformation in binary TeO_2-PbO glasses. In the case of TeO_2-PbO-GeO_2 glasses prepared in this study, the ratio of [TeO_4]:[TeO_3] remained nearly constant. Therefore, the decrease in $\chi^{(3)}$ values of TeO_2-PbO-GeO_2 glasses with increasing GeO_2 concentration was simply attributed to the decrease in the relative amount of [TeO_n] polyhedra and Pb^{2+} ions.

Increasing amount of [GeO_4] polyhedra, which normally show low polarizabilities compared to [TeO_n] and Pb^{2+}, inevitably resulted in the decrease of $\chi^{(3)}$ values even though it provided a remarkable improvement in the thermal properties.

CONCLUSION

In TeO_2-PbO glasses $\chi^{(3)}$ values are more dependent on PbO concentration than [TeO_4]tbp → [TeO_3]tp transformations. In TeO_2-PbO-GeO_2 glasses addition of GeO_2 resulted in the decrease of $\chi^{(3)}$ values. Measured $\chi^{(3)}$ values of 10.7-4.95$\times 10^{-13}$esu for these glasses were approximately 40-20 times of that of silica glass. Analyses of the Raman spectra suggested that TeO_2-PbO-GeO_2 glasses were mainly composed of [TeO_4], [TeO_3] and [GeO_4] structural units with [TeO_4]:[TeO_3] ratio being nearly constant. Decrease in $\chi^{(3)}$ values with increasing GeO_2 concentration was attributed to the rapid decrease in the amount of [TeO_n] polyhedra and Pb^{2+} ions due to the formation of [GeO_4] polyhedra with low polarizability.

ACKNOWLEDGMENTS

Authors would like to extend a special appreciation to Prof. H. Nasu of Mie University, Japan for his stimulating discussion and THG measurements.

REFERENCES

[1] E. M. Vogel, "Glasses as Nonlinear Photonic Materials", J. Am. Ceram. Soc., 72 [5], 719-724 (1989).
[2] H. Nasu, T. Uchigaki, K. Kamiya and H. Kubodera, "Non-resonant Type Third-Order Nonlinearity of (PbO, Nb_2O_5)-TiO_2-TeO_2 Glasses Measured by Third-Harmonic Generation", Jpn. J. Appl. Phys., 31, 3899-3900 (1992).
[3] S. H. Kim, T. Yoko and S. Sakka, "Nonlinear Optical Properties of TeO_2-Based Glasses: La_2O_3-TeO_2 Binary Glasses", J. Am. Ceram. Soc., 75 [4], 865-869 (1993).
[4] H. Kobayashi and K. Kubodera, "Analysis of Asymmetric Fringe Patterns of Third-Harmonic Generation in a Molecular Crystal", J. Appl. Phys., 69 [7], 3807-3810 (1991).
[5] T. Sekiya, N. Mochida, A. Ohtsuka and M. Tonokawa, "Raman Spectra of $MO_{1/2}$-TeO_2 (M=Li, Na, K, Rb, Cs, and Tl) Glasses", J. Non-Cryst. Solids, 144, 128-144 (1992).
[6] H. Verweij and J. H. J. M. Buster, "The Structure of Lithium, Sodium and Potassium Germanate Glasses, Studied by Raman Scattering", J. Non-Cryst. Solids, 34, 81-99 (1979).
[7] H. Yamamoto, K. Kamiya, J. Matsuoka and H. Nasu, "EXAFS Study on the Coordination State of Germanium in PbO-GeO_2 Glasses", J. Ceram. Soc. Jpn., 101 [9], 974-979 (1993).
[8] H. Yamamoto, H. Nasu, J. Matsuoka and K. Kamiya, "X-ray Absorption Fine Structure(XAFS) Study on the Coordination of Te in PbO-TiO_2-TeO_2 Glasses with High Third-Order Optical Non-linearity", J. Non-Cryst. Solids, 170, 87-96 (1994).
[9] S. H. Kim, T. Yoko, F. Miyaji, H. Kozuka and S. Sakka, "Third Harmonic Generation in PbO-TeO_2 and PbX_2(X=F, Cl, Br, I)-TeO_2 Glasses", in: Abstracts Annual Meeting of the Ceramic Society of Japan, Tokyo(Ceramic Society of Japan, Tokyo, 1992), p231.

IMPROVED PR^{3+}-DOPED HEAVY METAL FLUORIDE GLASSES AND FIBERS

James Fajardo and George H. Sigel, Jr.
Fiber Optics Materials Research Program, Rutgers University
Piscataway, NJ 08855-909

Praseodymium doped heavy metal fluoride glass preforms were fabricated via a modified suction casting method. A subsequent redraw process was employed to obtain high quality single mode fluoride fiber. Modifications of the ZBLAN composition, including the incorporation of In^{3+}, Y^{3+} and Cs^{+}, were undertaken in an attempt to improve optical properties while maintaining the inherent drawability of ZBLAN. Thermal analysis to determine the various draw parameters was conducted using Differential Scanning Calorimetry, while optical characterization included spectral absorption, fluorescence, and fluorescent lifetime measurements for all specimens. The modifications in composition resulted in an increase in the fluorescent lifetime of Pr^{3+} at 1.3μm.

INTRODUCTION

Pr^{3+}-doped fluoride fiber is an attractive candidate as a fiber amplifier at the 1.3μm telecommunications window, but there are some problems which must be overcome. A low concentration of Pr^{3+} in ZBLAN (<560ppmw PrF_3 in 53%ZrF_4-20%BaF_2-4%LaF_3-3%AlF_3-20%NaF) is necessary to avoid concentration quenching, which is attributed to non-radiative cross relaxation between neighboring active ions. [1] Because of this low concentration, longer lengths of fiber are required to obtain high gain. Although many techniques have been tried for the fabrication of high quality preforms, fluoride preforms made through the suction method have the advantages of a clean core-clad interface and the capability of being drawn into single-mode fiber. [2-4] One drawback, however, is that this approach produces a short, highly tapered preform with a non-uniform core to clad ratio.

Another issue to be addressed is the 110μsec fluorescent lifetime of the $^1G_4$ level for Pr^{3+} in ZBLAN, which leads to a low quantum efficiency of about 3.4%. A large non-radiative decay probability, which is dependent on the highest-energy phonons in the glass system, reduces the lifetime and quantum efficiency. [5] Since the frequency of these highest-energy vibrations decreases as the reduced mass of oscillating atoms increases, the incorporation of heavier elements into the glass typically results in a higher order multiphonon relaxation process, leading to an increase in the fluorescent lifetime. This paper reports lifetime data for Pr^{3+} in some glass compositions containing heavier fluoride constituents.

EXPERIMENTAL

Ultra-high purity starting materials were obtained from EM Fluortran for the fabrication of the fluoride glass preforms. All work was conducted in a controlled atmosphere glove-box using a furnace with reactive atmosphere processing (RAP) capability. Melting was done in vitreous carbon crucibles, and all melts were exposed to a N2:SF6 10:1 atmosphere at 750°C. The casting of the preform was by a modified suction casting process shown in Figure 1. The ZBLANPb core (2%PbF_2 added for index change) was poured over the ZBLAN clad, and the bottom latch of the mold was pulled after 40 seconds to allow the core composition to drop. This preform was drawn to 750µm diameter, which was then jacketed with a ZBLAN tube. The two were then drawn down to 150µm fiber to obtain single-mode. Loss measurements were conducted by a cutback loss method at two wavelengths, 632.8nm and 1.0µm, and cutback lengths of 2.5m-1m and 8m-1m were taken for testing.

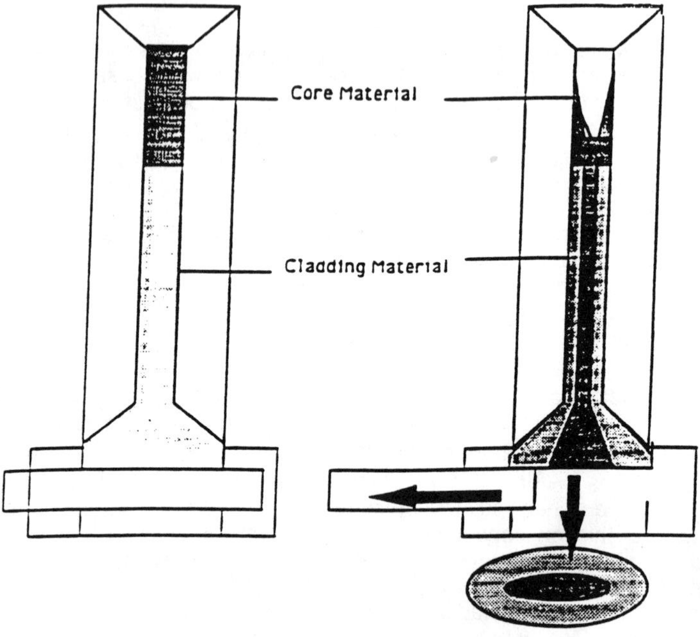

Figure 1: A Modified Suction Casting Method for Fabrication of Fluoride Preforms

The modifications in the ZBLAN composition included both adjustments in the percentages of the constituents (from 53-20-4-3-20) and substitution for AlF_3 and NaF with heavier compounds. Such compounds included YF_3, InF_3, GaF_3 and CsF. Differential scanning calorimetry was done on these samples from 100°C-500°C at 10°C/min.

Bulk samples of selected compositions were fabricated for the evaluation of optical properties. Absorption spectra were obtained through a Perkin Elmer Lambda 9 spectrophotometer from 190-3200nm at 60nm/min. Fluorescence and fluorescent lifetimes were measured using a Ti-sapphire laser pumped by an argon laser, and an InGaAs detector was positioned after a blazed grating monochromator for spectral data acquisition. Samples were pumped with 250mW at a wavelength of 1.01µm.

DISCUSSION

The utilization of this preform casting method resulted in a clad to core ratio of 3:1, which was consistent over a length of 70mm. The final fiber (after redraw) was single-mode with NA = 0.15 and a 4.5μm core. Cutback measurements showed a 0.6 dB/m background loss through a 1.5m length (2.5m-1m cutback). This was observed at both 632.8nm and 1.0μm, and thus is attributed to wavelength independent extrinsic loss. The 8m-1m cutback, however, resulted in a greater attenuation, 3.3dB/m, indicating that loss may be due to large scattering centers such as bubbles or small crystallites.

Table I: Thermal Properties of Various Fluorozirconate Glasses

composition	T_x (°C)	T_g (°C)	ΔT (°C)
ZBLAN	360	262	98
55-30-3-4-8	370	288	82
5% Cs	353	256	97
2% In	346	261	85
2% Y, 1% In	343	260	83
2% Ga, 1% In	338	253	85
2% Ga, 2% In	325	256	68
2% Y, 2% In, 5% Cs	326	258	68
2% Y, 1% In, 5% Cs	340	260	80

Selected compositions from the many modified ZBLAN compositions tested are listed in Table I. Also shown are the onset of crystallization temperature and the glass transition obtained through DSC. Although ΔT is not a true measure of thermal stability, it represents the working range where a preform can be drawn into fiber. As expected, the reduction in NaF from 20% to 8% increased both T_x and T_g towards values indicative of ZBLA. The incorporation of cesium for sodium did not have a great effect on ΔT, showing that such a substitution may not decrease the drawability of the glass. Conversely, ΔT dropped substantially for the Ga^{3+}, In^{3+} and Y^{3+}, In^{3+}, Cs^{+} systems when the indium content was increased from 1% to 2%.

Since the objective was to incorporate heavy elements without loss of drawability, optical properties were only studied for compositions where ΔT>80°C. Fluorescence spectra for Pr^{3+} in both ZBLAN and ZBLYINCs are shown in Figure 2. The 1.3μm emission peak characteristics for these and all other samples were similar. The $^1G_4$ lifetime of 500ppm Pr^{3+} was substantially increased, however, as the glass system included these heavier elements. The highest first e-folding time observed was 156μsec for the composition of ZBLYINCs (53-20-4-2-1-15-5). This improvement from 110μsec for ZBLAN, along with the fiber draw working range (T_x-T_g) of about 80°C, gives strong indication that Pr^{3+}-doped ZBLYINCs fluoride glass may be drawn into high quality single-mode fiber yielding higher gain at 1.3μm. Further compositional studies are continuing in an effort to optimize both fiber and amplifier characteristics.

Table II: Fluorescent Lifetimes of $^1G_4$ Level of 500ppm Pr^{3+} (in μsec)

composition	e^1	e^2	e^3
ZBLAN	109	172	312
55-30-3-4-8	129	174	313
2% In	117	152	279
2% Ga, 1% In	131	168	301
2% Y, 1% In, 1% Cs	140	180	277
5% Cs	150	230	328
2% Y, 1% In, 5% Cs	156	229	311

* 500ppm PrF_3, in μsec

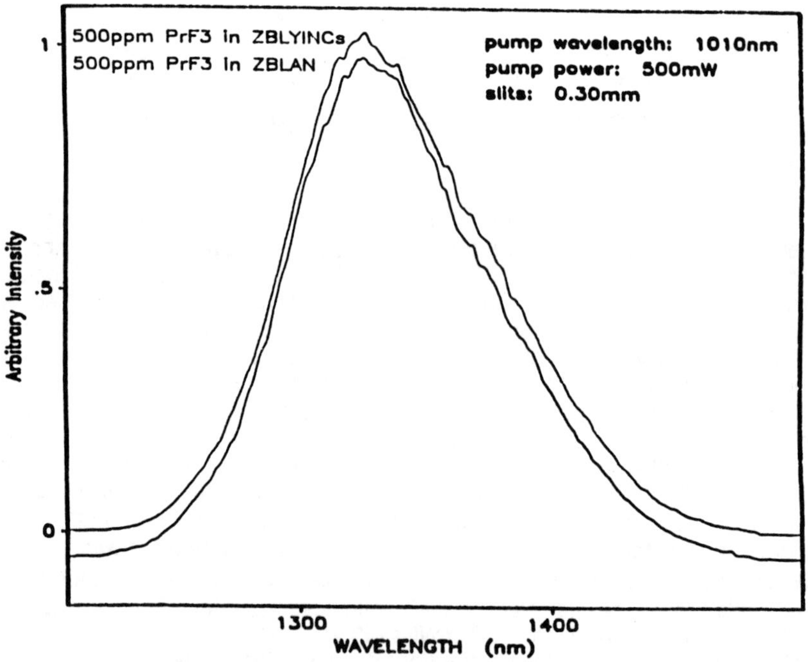

Figure 2: Fluorescence Spectra of Pr^{3+}

REFERENCES

1. S. F. Carter, D. Szebesta, S. T. Davey, R. Wyatt, M. C. Brierley, P. W. France, *Amplification at 1.3μm in a Pr^{3+} doped Single-mode Fluorozirconate Fibre*, Electronic Letters, 27, 3, pp 628-629, 1991
2. J. S. Sanghera, B. B. Harbison, I. D. Aggarwal, *Challenges in Obtaining Low Loss Fluoride Glass Fibers*, Jour. Non. Cryst. Solids, 140, pp146-149, 1992
3. H. W. Schneider and A. Staudt, *Influence of Various Casting Techniques on Optical Losses in Fluoride Glass Fibers*, Mat. Sci. Forum, vol. 67 &68, pp 303-316, 1991
4. Y. Ohishi, S. Sakaguchi, S. Takahashi, *Transmission Loss Characteristics of Fluoride Glass Single-mode Fibre*, Electronics Letters, 22, 20, pp 1034-1035, 1986
5. C. B. Layne, W. H. Lowdermilk, M. J. Weber, *Multiphonon Relaxation of Rare-earth Ions in Oxide Glasses*, Phys. Review B, 16, 1, pp 10-20, 1977

Er^{3+} - DOPED SODA-LIME GLASSES AND PLANAR WAVEGUIDE DEVICES.

A.J. Bruce, W.H. Grodkiewicz, J. Shmulovich, Y-H. Wong, P.C. Becker,
G. Nykolak, M. Barros and R.N. Ghosh.
AT&T Bell Laboratories, Murray Hill, NJ 07974, USA.

ABSTRACT.

The fabrication and optical performance of planar optical waveguide amplifiers, comprising of deposited Er^{3+}- doped soda-lime glass cores on silicon optical bench platforms is documented in this paper. The results are encouraging for practical device applications. A laser device has also been characterized. Its modest performance highlights some of the differences between amplifiers and lasers.

INTRODUCTION.

The advent of all-optical telecommunication networks through the introduction of Erbium doped optical fiber amplifiers (OFAs) has dramatically increased the data handling capacity of long haul systems in recent years [1]. Similar enhancements can be obtained in local area networks (LANs) operating in the 1.5 microns telecommunication band and are certainly desirable as fiber to the home/curb is deployed and telecommunication links are increasingly used for multi-media transmission.

At some point it becomes less attractive on the basis of cost, technology and/or reliability to interconnect the LAN's with individual fiber piece parts. This is the potential niche of integrated optical components in which multiple functionality is combined in single components [2]. In this context we are investigating the feasibility of planar optical waveguide amplifiers (POWAs) for silicon optical bench (SiOB) platforms. Recent results for prototype devices are encouraging and are presented in this paper.

POWA MATERIALS, FABRICATION AND MEASUREMENTS.

Typical OFAs are fabricated from 10-20 meters of high silica fibers which are lightly doped (< 200 ppm) with Er^{3+}. At higher dopant levels the Er ions tend to "cluster" and the performance of the OFA is compromised via co-operative upconversion and other energy loss mechanisms. POWAs on the other hand should be much shorter in length (<10-20 cm) and have a much higher level of Er^{3+} to achieve similar net concentrations of Er along the length of a device. For this application we are investigating multicomponent silicate glasses, including soda-lime glasses, as potential host materials for the Er ions. It is well known that lanthanide ions are much more soluble, and therefore should have a lower tendency to cluster, in such glasses [3]. Indeed, as shown in figure 1, measurements of peak radiative lifetimes as a function of 980 nm pump power for Er doped high silica fibers and soda-lime waveguides show that longer lifetimes can be maintained in soda-lime hosts at similar dopant levels. This is consistent with a lower energy loss through Er-Er cooperative effects [4].

As shown in figure 2 the emission profile of Er^{3+} in a soda-lime host is considerably narrower than in high silica. This may limit the application of soda-lime based devices for broadband amplification and dense wavelength division multiplexing (DWDM). However sufficient spectral broadening may be achieved by compositional modification of the host glass to address this issue [3].

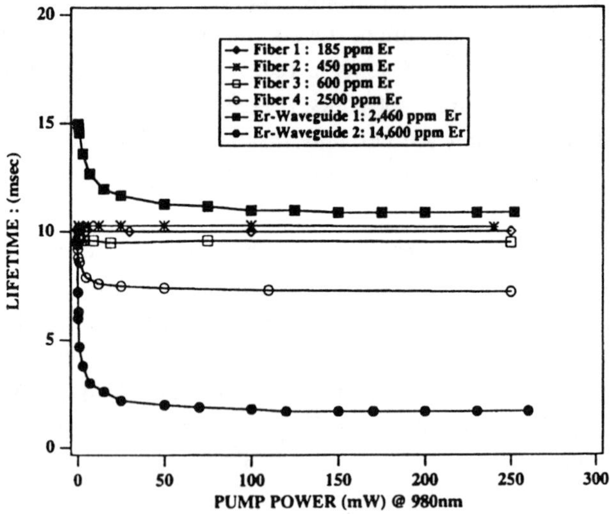

Figure 1. Radiative lifetime vs. pump power for fibers and waveguides.

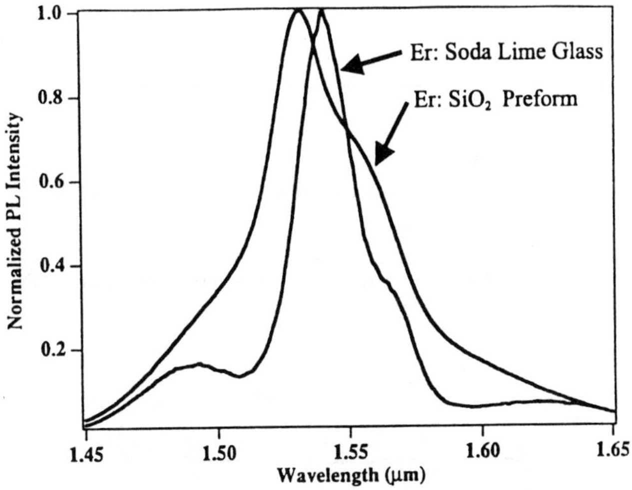

Figure 2 Emission spectra of Er in soda-lime and high silica glasses

The active core of our POWA devices were deposited by RF magnetron sputtering from bulk glass targets. The targets were prepared by conventional melting procedures and the sputtering process closely reproduces this target composition. Films up to 1.5 microns in thickness were deposited on Si wafers which had previously been oxidized under high pressure steam to produce a SiO_2 layer (HiPOx) approximately 15 microns in thickness. The HiPOx subsequently serves as a lower optical cladding. After deposition the films were annealed at 650°C to achieve densification. The waveguide structures were subsequently defined by standard photo-lithography procedures and ion milling. Typically the structures were slightly overmilled to insure isolation of the Er doped glass core. A schematic cross section is shown in figure 3. After definition the wafers were diced and the ends polished for subsequent optical evaluation.

The optical measurement system is shown in figure 4. A 1537 nm signal pulse is combined with a 980 nm pump and launched into the waveguide through an actively aligned, butt coupled, fiber which was tapered to provide a similar core dimension to the waveguide. The outgoing signal is similarly captured by a butt coupled fiber and sent to an optical spectrum analyzer. For these measurements a 1.458 index (n_d)oil was used as upper cladding. The relatively high index (n_d) of the core glass (1.51) also promotes good optical confinement.

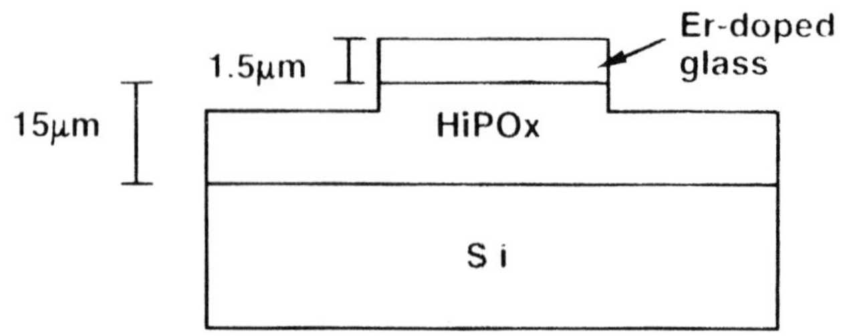

Figure 3 Cross section of Er-doped soda-lime POWA

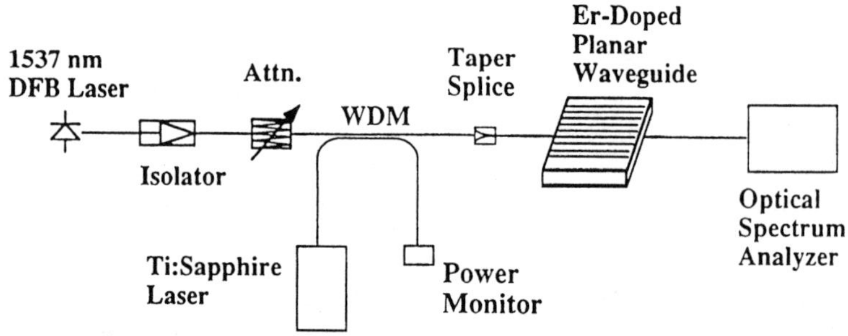

Figure 4 Optical measurement system for POWA.

RESULTS

Figure 5 shows the net gain as a function of 980 nm pump power for a 4.5 cm long soda lime waveguide containing 2.1 x 10^{20} Er ions cm^{-3}. The device demonstrates 4.2 dB/cm of gain at 375 mW with a threshold of approximately 40 mW. At 100 mW of pump power, a level commensurate with diode pumping, approximately 10dB of gain can be achieved. This result is already within the useful range for a number of proposed POWA applications.

A significant factor for achieving the above result was the development of waveguide reflowing. This is illustrated in figure 6. After ion milling the side walls of the waveguide core are rough. This contributes to propagation loss through scattering. We found that the sidewalls could be reflowed to a smooth surface by heating the patterned waveguide for 1.5 minutes at 850°C in a rapid thermal anneal (RTA) furnace. Reflowing reduces propagation losses from approximately 1.0 to 0.2 dB/cm. This effectively shifts the net gain spectrum higher and the threshold lower.

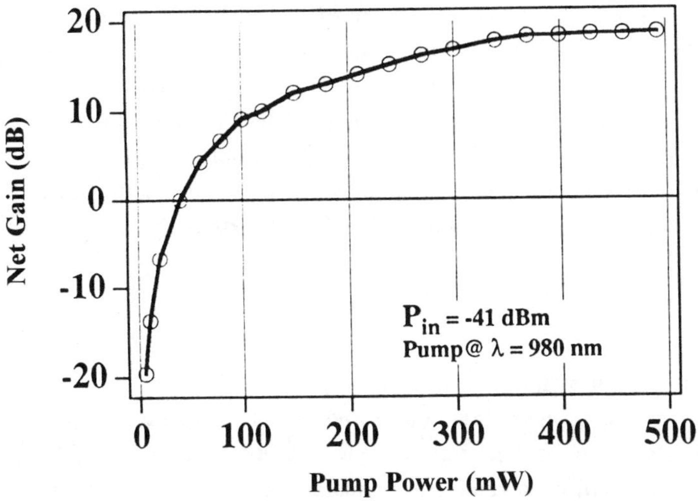

Figure 5 Gain characteristics for a 4.5 cm long soda-lime POWA.

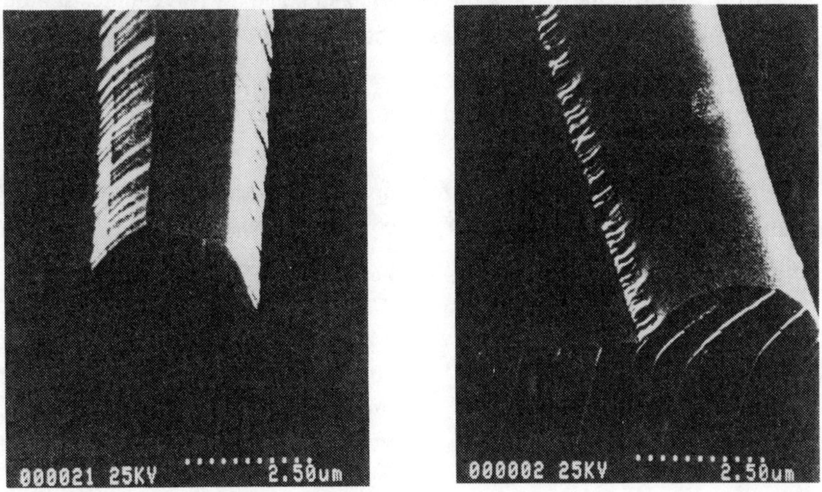

Figure 6 POWA after ion milling (left) after 850°C for 1.5 min (right)

PLANAR WAVEGUIDE LASER

To make a laser, a cavity was defined by writing gratings in the GeO_2 doped fibers which were butt coupled to a soda- lime waveguide (2.1 x 10^{20} Er ions cm^{-3}) as shown in figure 7. The 980 nm pump light is internally reflected within the cavity while the 1.5 micron emitted light is transmitted. The laser performance is very modest. The threshold was 100mW with a slope efficiency of 0.1%. Major factors in this low efficiency is the propagation loss of the waveguide (0.2 dB/cm) and the fiber to waveguide coupling loss (1dB/interface). These are encountered on every pass of the pump light and are cumulatively catastrophic for performance. This result illustrates the difference in acceptable loss for POWA and lasers. Similar waveguides can be excellent as POWA devices but be extremely poor lasers.

ISSUES.

While the optical performance of our prototype POWA device are encouraging significant issues must be addressed before they can be fully integrated with other SiOB components. Many of these pertain to the compatibility of the materials and processing with standard waveguide components. e.g.

(i) A low index upper cladding must be developed which is fully processed at temperatures commensurate with the soda lime cores.

(ii) A low loss transition between the POWA and standard waveguides should be developed. Mismatched geometries, dimensions and index result in significant coupling losses.

(iii) Gain spectra should be broadened by compositional modification to permit DWDM operation.

(iv) The gain threshold should be decreased further by optimizing processing to reduce propagation loss in the Er doped waveguides.

(v) Integration and packaging strategies have to be developed.

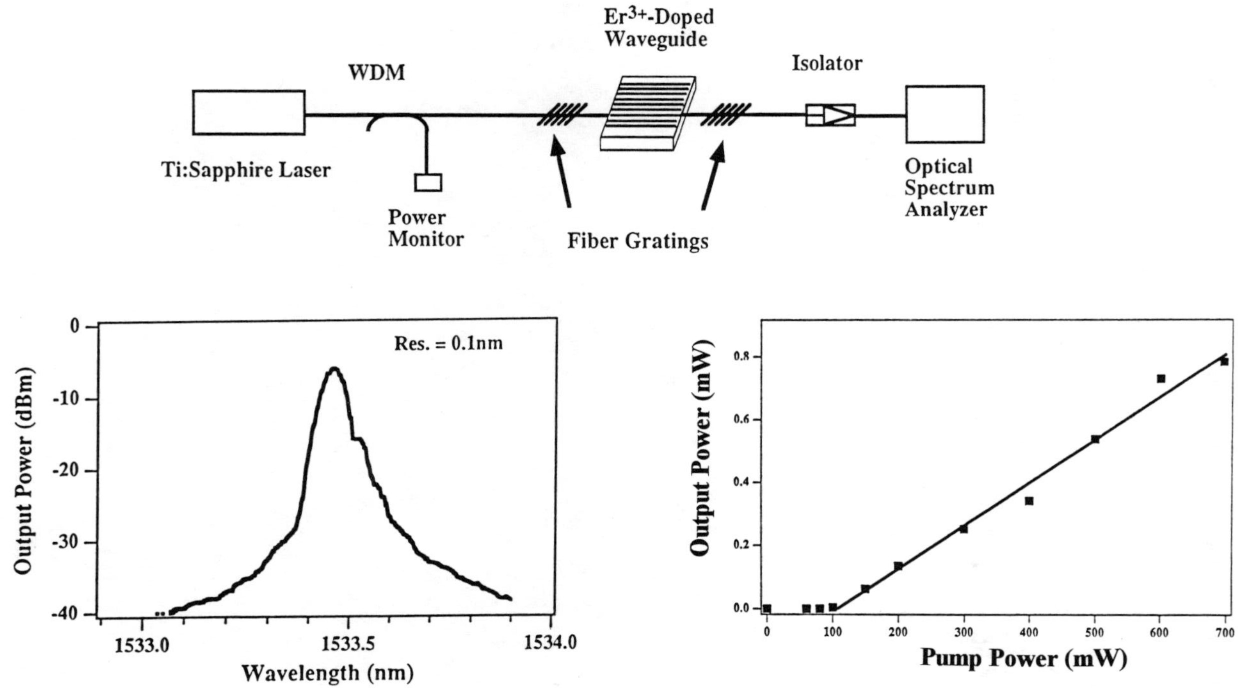

Figure 7 Er doped soda-lime waveguide laser and performance characteristics

REFERENCES

1. E. Desurvire, Erbium Doped Fiber Amplifiers, John Wiley, New York, 1994.
2. C. H. Henry, Proceedings Australian Conference on Optical Fiber Technology, (ACOFT ' 94), Melbourne , Australia, December 1994.
3. A.E. Neeves, A.J. Bruce, W.A. Reed, E.M. Rabinovich, W.H. Grodkiewicz, N.A. Kopolov, A. Lidgard and D.DiGiovanni, Ceram. Trans. vol 28, 353,1992
4. G. Nykolak, P.C. Becker, J. Shmulovich, Y-H. Wong, D.J. DiGiovanni and A.J. Bruce, Photonics Technology Lett., 5, 1014, 1993.

Phosphors and Scintillators

PHOSPHORS FOR FLAT PANEL EMISSIVE DISPLAYS

Mark T. Anderson, Robert J. Walko, and Mark L. F. Phillips, Sandia National Laboratories, Albuquerque, NM 87185-5800.

ABSTRACT

An overview of emissive display technologies is presented. Display types briefly described include: cathode ray tubes, field emission displays, electroluminescent displays, and plasma display panels. The critical role of phosphors in further development of the latter three flat panel emissive display technologies is outlined. The need for stable, efficient red, green, and blue phosphors for RGB full color displays is emphasized.

INTRODUCTION

A. DISPLAYS IN THE INFORMATION AGE

Technology for flat panel displays (FPDs) has evolved rapidly in the past several years as demand has continued to increase.[1] In 1992 Hitachi predicted that the burgeoning market for displays would exceed $40 billion by the year 2000. Displays for video applications can be grouped arbitrarily into three general sizes: small (up to 14" diagonal), medium (up to 35" diagonal), and large (greater than 35" diagonal). We shall not discuss small nonvideo displays that rely on vacuum fluorescent, liquid crystal, or light emitting diode technologies such as those used for alphanumeric indictors. Small video displays are used, for example, as monitors for laptop computers and as monitors for military displays. For portable, battery operated displays, the devices must be lightweight, thin, and efficient. Liquid crystal displays (LCDs) currently hold 90 to 95% of the market for small flat panel displays. The flat panel emissive display technologies that could challenge this technology are field emission displays (FEDs) and electroluminescent displays (ELDs). Medium size displays are employed commonly as television sets and as desktop computer monitors. Owing to their relatively low cost, stable performance, and good visual output, cathode ray tubes (CRTs) are the dominant technology. It would be advantageous to replace these bulky monitors with thinner, lighter, more portable flat panel displays. The FED,

This work was funded by the United States Department of Energy Under Contract No. DE-AC04-94AL85000.

ELD and plasma display panel (PDP) are potential challengers along with LCDs if technical barriers to size scale up can be overcome. Large displays are used typically in big venues, such as, military command centers, meeting halls, public facilities, and classrooms. Projection displays based on CRTs and LCDs are the most common technologies. Monochrome PDPs have been available for over 20 years and have found niche markets in medical, commercial, and military displays.[2] Desirable attributes of new large displays include full color, high resolution (for use as direct view high definition televisions, HDTVs), and readability in bright light. Phosphors are one key to further development of flat panel emissive display technologies. The properties and performance of phosphors exhibit a wide variability. The choice of phosphor type and deposition method provides a great opportunity to improve the performance of a display.

B. PHOSPHORS: KEYS TO EMISSIVE DISPLAYS

Phosphors are materials that emit visible light when excited by photons or electrons. It is convenient to describe phosphors in terms of a host, activator, and coactivator. Phosphor hosts are typically crystalline oxides or sulfides. Activators are typically lanthanide ions, transition metal ions, or molecular complexes. Coactivators are ions that increase luminescent efficiency of the activator. Phosphors are generally sprayed, shaken, or screen printed in powder form onto the faceplate of a display in small dots or stripes (pixels) that are on the order of 0.3 mm for CRT applications. For some applications they are deposited as thin films. In general full color displays contain red, green, and blue (RGB) pixels. The use of RGB in full color applications dictates that three separate materials, all of which meet the necessary criteria for operation in a display, must be developed for each full color display technology. Important figures of merit for traditional applications of phosphors (in CRTs) include:[3] chromaticity (color) specified as CIE coordinates, peak wavelength of visible emission, luminous efficiency (W/W or lm/W), persistence (decay time for the light output after the input energy is removed), and maintenance (luminance of phosphor relative to that when the screen was first excited). Another important benchmark is luminance. A standard CRT produces a maximum luminance of approximately 100 foot Lamberts (fL) for white light. The white luminance is composed of 10 fL blue, 25 fL red, and 65 fL green. The effeciency with which the light is produced is also an important metric. For portable displays the efficiency needs to be 2-3 lm/W to compete with LCDs. This requires 1 lm/W blue, 2-3 lm/W red, and 6 lm/W green from the phosphors.

C. PHOSPHORS IN CATHODE RAY TUBES

The excitation mechanism of phosphors in a CRT is cathodoluminescence, that is, high energy electrons strike the phosphors and excite the activators, which radiatively decay and emit visible light. A color CRT contains three thermionic electron guns. The electrons produced by the guns are rastered across the phosphor screen. The applied voltage is typically 10 to 30 kV. Phosphors in powder form are used in CRTs. The red, green, and blue phosphors are sequentially deposited onto a faceplate. The area for each phosphor is defined by photolithography. A shadow mask in conjunction with a light in the position where each electron gun will be placed in the display are used to expose the photoresist and define the pixels. For CRT phosphors the host lattice is a generally metal sulfide or metal oxysulfide. Sulfides exhibit energy efficiencies in the range 22 to 25%, whereas the maximum for oxides is 10 to 11 %. ZnS doped with various activators is the most extensively investigated phosphor scheme. Phosphors for green light include (Zn,Cd)S:Cu,Al and ZnS:Cu,Au,Al. Phosphors for blue light are ZnS:Ag,Cl and ZnSAg,Al. ZnS-based phosphors for red light are no longer used owing to problems with oxidation and poor saturation properties. The standard red phosphor is now Y_2O_2S:Eu. Many other lanthanide-doped materials have been developed since the 1950s. These include oxysulfides and oxides, such as, La_2O_2S:Tb, $Y_3(Ga,Al)_5O_{12}$:Tb, Y_2SiO_5:Tb, $Y_3Al_5O_{12}$:Ce, Y_2SiO_5:Ce, Y_2O_3:Eu, and YVO_4:Eu.

FLAT PANEL EMISSIVE DISPLAYS

A. FIELD EMISSION DISPLAYS (FEDs)

FEDs are analogous to CRTs. Both are vacuum technologies in which the phosphor excitation mechanism is cathodoluminescence. FEDs are operated at lower voltages and higher current densities than CRTs. A schematic of an FED is shown in Figure 1. One version of an FED relies on an addressable array of cold cathode field emitters[4] to provide electron emission. The cathodes are typically arrays of molybdenum or silicon points[5] or wedge emitters that are fabricated by the use of integrated circuit technology. A density as high as 10^8 emitters per square centimeter can be achieved. Another version relies on emission of electrons from diamond-like thin films.[6] Much of the early work on FEDs can be attributed to LETI in France. Prototype FEDs were demonstrated as early as 1991.[7]

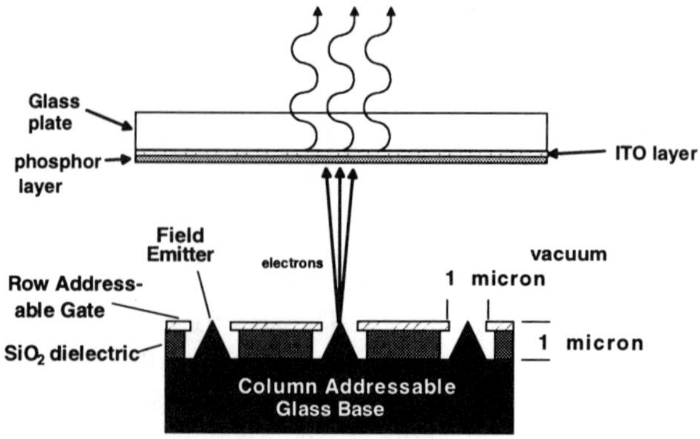

Figure 1. Schematic of a field emission display. Spacers between emitters and faceplate are not shown.

There are a variety of operational voltages being targeted for use in FEDs. The voltages range from 5 kV to less than 500 V. Each voltage range has distinct advantages and disadvantages. Low-voltage operation is probably more viable, especially if reliable, efficient low voltage phosphors can be developed.

Low-voltage operation, especially at less than 500 V, is attractive from a safety standpoint and less stringent requirements to avoid vacuum breakdown. Low-voltage operation requires higher current densities to maintain the same average power at the phosphor, which can lead to increased degradation and outgassing of the phosphor.[8] Low-voltage operation also requires phosphors that have low threshold voltages, high thermal conductivity, and requires electrical conductivity of the phosphor/binder matrix to dissipate charge. A conductive layer, such as indium tin oxide, on the faceplate will likely be necessary to help prevent charging of the phosphor layer. Another drawback of low-voltage operation is that phosphor efficiency decreases as voltage decreases at a fixed input power. An efficiency of 2-3 lm/W white is desirable for the device. Many phosphors do not achieve this at 500 V. Also low-voltage, high-current driver development would be required as the final power amplification is provided by the drivers rather than the phosphor screen.

High-voltage operation is attractive to overcome phosphor degradation and charging problems. In this case it will probably be possible to aluminize the phosphor screen as in a CRT and maintain the same efficiency of the phosphors as for low-voltage conditions with no aluminum screen. The aluminum screen provides a physical barrier between the phosphors and emitter tips that can potentially reduce the concentration of volatile contaminants at the emitters. It also provides electrical conductivity and reflects emitted visible light that radiates toward the aluminum screen back toward the viewer. High voltages introduce several major concerns, these include: increased possiblility of vacuum breakdown owing to larger field gradients, increased electrical leakage, potentially more damage to the emitters from an increase in sputtered contaminants from the screen owing to the increased energy of the impinging electrons, the potential for more damage to the device in failure situations, and increased need for aging and testing of the each device prior to shipping. Additionally, high-definition devices would require focused electron beams, which adds cost and complexity to the package.[9] Phosphors developed for CRTs may be adequate for use in high voltage FEDs (if they do not outgas or decompose).

The inherent challenges of high-voltage operation underscore the importance of low-voltage operation. Many of the problems with low-voltage operation can be overcome with improved phosphors. For low-voltage operation, ZnO:Zn shows the most promise for operation at 500 V or less. Reduced zinc oxide exhibits green-blue fluorescence and can be excited by electrons that have energies as low as 3 eV. It is reported to deliver a luminous efficiency of 13 lm/W at 5 V, and exhibit good operational lifetime under high current densities;[10] however, the green-blue color is a problem in full color applications. Suitable chemical dopants that produce other colors efficiently have not been found.[11] Table I[12,13,14] shows a variety of phosphors examined at 500 V.

Table I. Composition, Color, and Efficiency of Commercial and Experimental Low-Voltage Phosphors

material	color	efficiency at 500 V (lm/W)	ref
ZnS:Ag	blue	0.8	12
$ZnGa_2O_4$	blue	0.2	12
ZnO:Zn	green	10.7	12
Gd_2O_2S:Tb	green	7.9	12
(Zn,Cd)S:Cu,Al	green	4.8	12
$ZnGa_2O_4$:Mn	green	2.2	12
$Y_3Al_5O_{12}$:Tb	green	2.0	13
Y_2O_2S:Eu	red	3.5	12
Y_2O_3:Eu	red	2.2	12
YVO_4	red	0.4	14

As indicated in Table I the traditional blue phosphor for CRTs, ZnS:Ag, and one of the more promising oxide phosphors, $ZnGa_2O_4$, both have efficiencies less than 1 lm/W. There is a report that the blue emitter zinc gallate has a threshold voltage of 8 V and a luminous efficiency of 0.7 lm/W when the display is operated at 30 V DC.[15] Results in Table I do not confirm this luminous efficiency at low voltages. For red phosphors the La_2O_2S:Eu appears to have acceptable efficiency, but long term stability under high current densities is a potential problem. La_2O_3:Eu is an alternative, but the efficiency at 500 V is marginal and the red color saturation is worse than for La_2O_2S:Eu.

B. ELECTROLUMINESCENT DISPLAYS (ELDs)

In ELDs, high energy electrons from the insulator layer are accelerated into the phosphor layer where they excite the activators; the activators radiatively decay to produce visible light. ELDs are commonly made in thin film form, or TFEL. A schematic of a TFEL display is shown in Figure 2. Typical voltages across the electrodes are 100 to 300 V. Typical AC frequencies employed range from 60 Hz to 1kHz. DC devices can be made in powdered form and rely on a copper sulfide ZnS heterojuction;[16] however, the devices frequently have limited operational lifetimes.

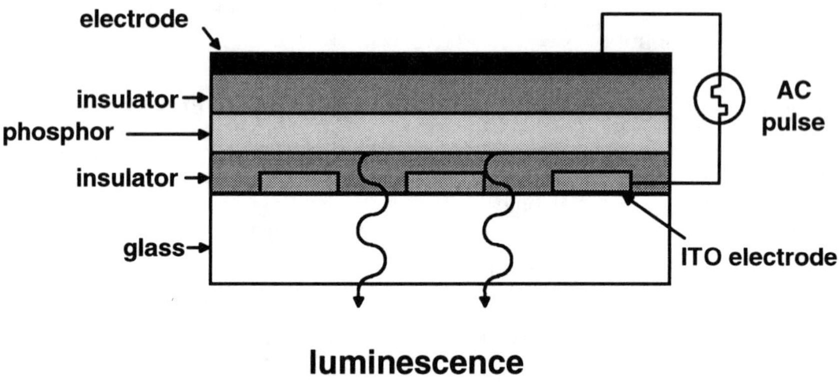

Figure 2. Schematic of an AC thin film electroluminescent display.

Phosphors for TFEL to be used in FPDs must be able to produce desired luminance, have efficiencies of 2-3 lm/W, have threshold voltages well below the operational voltages, be stable in the voltage field, be chemically stable, have low capacitance and resistance, and be amenable to thin film deposition techniques. The brightest and most efficient phosphor material for these devices is ZnS:Mn.[15] A yellow luminescence is produced that provides luminance up to 120 fL with efficiencies of 5 lm/W.[17] Commercial monochrome devices based on ZnS:Mn phosphors have been available for many years. An example is a 17" diagonal monochrome monitor that exhibits 65 fL with VGA resolution is available from Sharp.[18]

Red, green, and blue phosphors that have luminances that exceed the desired values stated eariler (25, 65, 10 fL) have not been produced to date, although many studies have been performed in pursuit of improved materials and deposition techniques.[19,20,21,22,23] Colors other than yellow can be produced by doping ZnS with rare earths.[24] Spire along with Planar and Norden have reported several phosphors in which the activator is ion implanted into the thin film.[17] The brightest is ZnS:Tb,[25] which emits green light and has a luminance of 10 fL at 180 Hz. The efficiency is less than 1 lm/W. In addition to ZnS, $CaGa_2S_4$ has promise as a host. It has been doped with Eu to produce green emission (2.5 fL), with Mn and Sm to produce a weak red emission, and with Ce to produce a blue emission (maximum 3 fL).[26] SrS:Ce[27] provides a greenish-blue output of 20 fL, but is hygroscopic. In all of the ion implanted phosphor films O, F, or S counterions were also implanted for charge balance.[17] Luminance is strongly dependent on counterion type and conditions used to anneal the samples.[17] Currently all of the full color devices fall short of the desired 100 fL luminance.

C. PLASMA DISPLAY PANELS

The excitation mechanism of the phosphor in PDPs is photoluminescence. The excitation principal is the same as in the original monochrome neon glow discharge devices. In the case of PDPs a noble gas mixture (typically 93 % He + 5% Kr + 2% Xe) is ionized when sufficient voltage is applied across addressable electrodes adjacent to the gas. The ionized gas molecules emit photons. The high energy photons from the discharge are used to stimulate visible emission from the phosphors. In the display it is necessary to incorporate barriers between each pixel to reduce cross talk between the phosphors. A schematic is shown in Figure 3. The simplest PDPs operate in DC mode. The electrodes are located within the gas container. The phosphors are deposited on or near the anode, which can lead

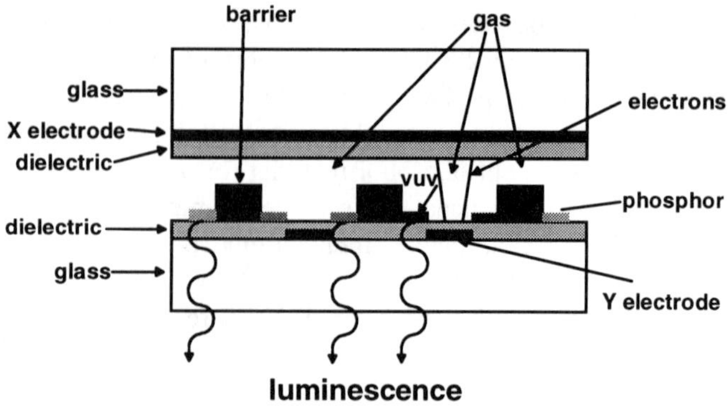

Figure 3. Schematic of a plasma display panel.

to contamination of the electrodes. Problems from contamination of the electrodes can be mitigated by the use of an AC driven design.[28] The column and row electrodes can be insulated from the gas discharge. AC PDPs can be driven at frequencies as high as 250 kHz. The high driver frequency helps to increase luminance.

Typical phosphors used in PDP are Zn_2SiO_4:Mn for green, Y_2O_3:Eu for red, and $BaMgAl_{14}O_{23}$:Eu for blue. The largest issues for the phosphors is their stability in the ionizing conditions in the display and their adherence.[29] Efficiency is less of a concern for these displays that run from fixed rather than portable power supplies.

PDPs are a rather mature technology. PDPs were introduced in the 1960s and have been applied in commercial, industrial, and military programs in significant volumes.[2] The largest area flat panel display of any kind is the 58.6" diagonal AC PDP built by Photonics Systems in 1986. Photonics Imaging demonstrated a 30 " full color AC-PDP video monitor in 1987.[30] Fujuitsu has a commercially available 21" full color AC-PDP that provides 58 fL with full color resolution of 640 x 480 and 256 level gray scale.[31] Thomson has demonstrated a 13" color AC-PDP with a pixel pitch of 0.4 mm, the highest resolution color reported to date.[32] Recently, NHK has announced the fabrication of a 40" DC PDP with a white areal luminescence of about 20 fL and a luminous efficiency of 0.15 lm/W. Trends are toward higher resolution, increased operational lifetime, and lower cost. There do not appear to be any fundamental impediments to fabrication of 50 or 60" diagonal displays.

SUMMARY OF OPPORTUNITIES FOR PHOSPHOR RESEARCH AND DEVELOPMENT

Many opportunities exist to improve phosphors for flat panel emissive displays. For FEDs the major issues are stability under high current densities and efficiency especially of the blue and red phosphors. For ELDs the challenges is to develop red, green, and blue phosphors that are stable and produce requisite luminance for monitor applications. For PDPs issues of adherence and stability must be addressed.

ACKNOWLEDGMENTS

We thank B. G. Potter, Jr. and the American Ceramics Society for their invitation. One of us (RJW) acknowledges the National Center for Advanced Information Component Manufacturing (NCAICM) at Sandia National Laboratories and the Phosphor Technology Center of Excellence (PTCOE). This work was funded by the United States Department of Energy under Contract No. DE-AC04-94AL85000.

REFERENCES

1. W. F. Goede, 1993 *SID* Seminar M-0, 3-28.
2. R. A. Stoller and J. F. Nolan, Proc. SPIE Vol. 2174, P.S. Friedman, Ed. 174-180.
3. see for example: Phosphor Resource Manual for Industrial and Military Cathode Ray Tubes Imaging and Sensing Technology Corporations.
4. R. F. Millikan and C. F. Eyring, *Phys Rev.* **27**, 51 (1926).
5. C. A. Spindt, *J Appl. Phys.* **39**, 3504 (1968).
6. M. W. Geis, N. N. Efremow, J. D. Woodhouse, M. D. McAleese, M. Marchywka, G. G. Socker, and J. F. Hochedez, *IEEE Electron Dev. Lett.* **12**, 456, (1991).
7. A. Ghis, R. Meyer, R. Rambaud, F. Levy, and T. Leroux, *IEEE Trans. Electron Devices* **ED-38**, 2320 (1991).
8. S. Itoh, T. Kimizuka, and T. Tonegava, *J. Electrochem. Soc.* **136**, 1819 (1989).
9. A. G. Chakhovskoi, W. D. Kesling, J. T. Trujillo, and C. E. Hunt, *J. Vac. Sci. Technol. B* **12(2)**, 785-789 (1994).
10. K. Narita, A. Kagami, and Y. Mimura, *J. Electrochem. Soc.: Solid-State Science and Technology* 1794-1798 (1980).

11. V. A. Nikitenko, S. V. Mukhin, S. P. Pivneva, and I. P. Kuz'mina, *Inorganic Materials* **29(10)**, 1253-1255 (1993).
12. S. Yang, et al. SI&T/SPIE Februrary 6-10, San Jose, CA, 1994.
13. M. L. F. Phillips, Ceramics Transactions (this issue).
14. M. L. F. Phillips, Proc. SPIE Vol. 2408, 200-206.
15. S. Itoh, H. Toki, Y. Sato, K. Morimoto, and T. Kishino, *J. Electrochem. Soc.* **138**(5), 1509-1512 (1991).
16. A. Zeinert, P. Benalloul, J. Benoit, C. Barthou, J. Dreyhsig, and J. E. Gumlich, *J. Appl. Phys.* **71**, 2855 (1992).
17. H. P. Maruska, T. Parodos, N. M. Kalakhoran, and W. D. Halverson, Mater. Res. Soc. Symp. Proc. Vol. 345, 1994; pp. 269-280.
18. K. Werner, *Information Display* **9**, 10 (1993).
19. R. Mach and G. O Muller, *Phys. Stat. Sol. A* **69**, 11 (1982).
20. V. P. Taninen, M Oikkonen, and T. Tuomi, *Thin Solid Films* **109**, 458 (1990).
21. P. J. Dean, *Phys. Stat. Sol A* **81**, 625 (1984).
22. V. V. Ushakov and A. A. Gipius, *J. Crystal Growth* **101**, 458 (1990).
23. P. A. Holloway, J.-E. Yu, P. Rack, J. Sebastian, S. Jones, T. Trottier, J. B. Pathangey, T. J. Anderson, S.-S. Sun, R. Tuenge, E. Dickey, and C. N. King, Mat. Res. Soc. Symp. Proc. Vol. 345, 1994; pp. 289-298.
24. M. K. Jayaraj and C. P. G. Vallabhan, *J. Electrochem. Soc.* **138**, 1512 (1991).
25. K. Okamoto, T. Yoshimi, and S. Miura, *Appl. Phys. Lett.* **53**, 678 (1988).
26. W. A. Barrow, R. C. Coovert, E. Dickey, C. N. King, C. Laakso, S. S. Sun, R. T. Tuenge, and R. Wetross, *SID 93 Digest*, 761-764 (1993).
27. S. Tanaka, H. Deguchi, Y. Mikami, M. Shiiki, and H. Kobayashi, SID Digest, 29-32 (1986).
28. T. Shinoda, Y. Miyashita, Y. Sugimoto, and K. Yoshikawa, 1981 SID Symposium, New York, 154-165.
29. C. H. Perry, Extended Abstracts, Electrochemical Society Meeting, Los Angeles, 783-784 (1979).
30. P. S. Friedman, T. J. Soper, T. D. Holloway, C. D. Reuter, and D. K. Wedding, Sr., Proc. SID, 28/4, 365-369 (1987).
31. T. Shinoda, et al. *SID 93 Digest* May, 161-164 (1993).
32. G. Baret, et al. *SID 93 Digest* May, 173-175 (1993).

PHOTO- AND CATHODOLUMINESCENCE OF HYDROTHERMALLY SYNTHESIZED $Y_3Al_5O_{12}$:Tb AND $NaY(WO_4)_2$:Tb

Mark L. F. Phillips and B. G. Potter, Jr.
Sandia National Laboratories, Albuquerque, NM 87185

ABSTRACT

Cathodoluminescent (CL) phosphors with improved low-voltage characteristics are needed for use in emissive flat panel displays. Conventional high-temperature methods for phosphor synthesis yield large polycrystalline grains that must be pulverized prior to screen deposition. Grinding has been implicated in reducing phosphor efficiency by causing surface contamination and defects. Hydrothermal synthesis has been used to improve the quality of ceramic powders by producing fine, well-formed crystallites without grinding. Two green-emitting phosphors, $Y_3Al_5O_{12}$:Tb (YAG:Tb) and $NaY(WO_4)_2$:Tb, were used to test the effects of hydrothermal synthesis on grain size and morphology, and on low-voltage CL properties. YAG:Tb prepared hydrothermally consisted of submicron crystallites with a typical garnet habit. The CL efficiency of hydrothermally synthesized YAG:Tb (3 lm/W at 800 V) was comparable to that of equivalent YAG:Tb compositions prepared via high-temperature solid state reaction. In comparison, CL intensities of $Gd_3Ga_5O_{12}$:Tb were slightly better (3.5 lm/W at 800 V), while those of $NaY(WO_4)_2$:Tb were approximately 1/100th that of YAG:Tb. Both CL and photoluminescence data show that the difference in the cathodoluminescence of YAG and $NaY(WO_4)_2$ can be understood in terms of differences in the mechanism of activation.

1. INTRODUCTION

Field emission-based displays (FEDs) are beginning to compete with liquid crystal displays (LCDs) for direct-view flat panel display applications. Full color FEDs are expected to offer several advantages over LCDs, including lower power consumption, lower manufacturing costs, and improved screen appearance through wider viewing angle and enhanced resolution and brightness. FEDs use electron beams to excite cathodoluminescent phosphors deposited on a screen, as conventional cathode ray tube (CRT) displays do. But while a CRT display rasters a beam from a single thermionic electron gun (or three guns, if in full color) across a phosphor screen, a FED uses an array of microscopic cold cathodes to irradiate the screen. Such arrays can contain as many as 10^7 emitters/cm^2, so that thousands of emitters can be used to illuminate a single pixel.[1]

FEDs also differ from CRTs in that they operate at lower voltages. The electron guns in CRT displays produce electrons at high energy (typically 10-25 keV). Because the cold cathodes in an FED must be held a short distance from the screen to obtain adequate resolution, they must be operated at lower voltages in order to avoid vacuum breakdown. In order to maintain adequate screen brightness, the electron current must be increased to compensate for the decreased voltage. Phosphors for FEDs must therefore not only be efficient at low voltages, but also resistant to saturation and Coulomb aging at high current densities.

In order to compete with LCDs in terms of power consumption, full-color FEDs will need a three-color phosphor scheme with a white brightness efficiency of 2-3 lm/W. With good color saturation, this requires green, red, and blue phosphors with efficiencies of approximately 6, 3, and 1 lm/W, respectively. These efficiencies can be achieved at 500 V using sulfide phosphors such as Y_2O_2S:Eu (red, 3.5 lm/W), Zn,CdS:Cu,Al (green, 4.8 lm/W), and ZnS:Ag (0.8 lm/W).[2] However, sulfide phosphors can release sulfur-containing gases under low-energy electron irradiance.[3] In addition to damaging the phosphor grain, this degradation can contaminate the field emitter tips, changing their work functions and decreasing (or catastrophically increasing) emission current.[4]

Oxide phosphors are less likely to contaminate the atmosphere within the FED, but few oxide CL phosphors are as efficient as the sulfides. Most oxide phosphors tend to be inefficient at low voltages (less than ca. 500 V), and most exhibit a threshold voltage below which no light is emitted. This can be as high as 200-300 V.[5] Using higher current densities to compensate for loss of efficiency at low excitation voltage can also lead to saturation and/or damage. Dead voltage is caused principally by surface bound electrons (SBEs) and surface recombination (SR) centers. SBEs result from secondary electron emission and cause the surface of the phosphor grain to become negatively charged; this surface charge repels or deflects the incident electron beam. SR centers are surface defects at which excitons created by electron impact can recombine nonradiatively. These defects can be caused by surface contamination or defects introduced by grinding.[6]

ZnO:Zn is unusual among oxide phosphors in that it is both highly efficient (13 lm/W) and electrically conductive, and is luminescent with electron energies as low as 5 eV.[7] However, ZnO:Zn is susceptible to saturation (efficiency loss) at moderate current densities, and the blue-green visible emission from this phosphor has poor chromaticity values. Rare-earth doped phosphors have the advantage of using higher activator concentrations, which allow them to be used at higher electron currents before saturating. Saturation resistance is a requirement in projection CRTs, which frequently use $Y_3Al_5O_{12}$:Tb (YAG:Tb) as a green-emitting phosphor.[8] The low-voltage cathodoluminescence of YAG:Tb and its isostructure $Gd_3Ga_5O_{12}$:Tb (GGG:Tb) are studied in this work. Another rare-earth host, $NaY(WO_4)_2$, was also studied as a CL phosphor. This phase is isostrucural with the fluorescent mineral scheelite, $CaWO_4$, and has been studied as an anti-Stokes phosphor when codoped with Yb and Er, Tb, or Tm, and as a laser host when doped with Nd.[9] The high photoluminescent (PL) efficiency of $NaY(WO_4)_2$:Tb and the similarity of its emission spectrum to that YAG:Tb warrant an investigation of its low-voltage CL properties.

Hydrothermal synthesis is a route to producing well-crystallized oxide powders with good control over particle size distribution. This process can yield fine particles without grinding, and thus may reduce surface defects, contamination, and threshold voltage. In addition, the very fine (1 μm or smaller) particle sizes available allow smaller pixel sizes, which are needed in high resolution displays such as projection CRTs. Phosphors and host phases that have been synthesized hydrothermally include YAG:Tb,[10,11] $ZnGa_2O_4$,[12] Zn_2SiO_4:Mn,[13] Y_2O_3,[14] and YVO_4:Eu.[15]

In this work both photoluminescence and low-voltage cathodoluminescence are studied in YAG:Tb and $NaY(WO_4)_2$:Tb phosphors. These materials were synthesized via both hydrothermal and conventional high-temperature solid state routes to determine whether hydrothermal synthesis affects CL efficiency, particularly at low voltages, and if this method can be used to obtain phosphor powders with finer grain sizes. The hydrothermal synthesis of $NaY(WO_4)_2$ and the CL properties of hydrothermally synthesized YAG:Tb are reported here for the first time. The mechanisms of activation in YAG:Tb and $NaY(WO_4)_2$:Tb and their effects on CL efficiency are also investigated and compared.

2. EXPERIMENTAL

2.1 Synthesis

$Y_3Al_5O_{12}$:Tb and $Gd_3Ga_5O_{12}$:Tb powders were prepared by high temperature solid state reaction:[16] Y, Tb, and Al oxide precursors were combined in stoichiometric proportion, pre-reacted at 1200 °C for 18 hr, reground, and annealed at 1450 °C for 6 hr. The YAG:Tb compositions were also synthesized hydrothermally using a method similar to that described in ref 10, starting with nitrate salts of Y, Al, and Tb. The Tb/Y ratio in the starting material was equal to that in the desired product, while Y and Tb were added in excess of the (Y:Tb)/Al product stoichometry. The excess rare earth was necessary to prevent formation of $\alpha-Al_2O_3$, since YAG saturates incongruently in water.[11] The Y, Al, and Tb salts were dissolved in water, then hydrolyzed by adding aqueous NH_3 until a pH of 10 was reached. The resulting gels were then filtered, dried at 150 °C, and heated at 500 °C for 6 hr in N_2. This yielded amorphous YAG:Tb precursors that were then ground, individually sealed in Au tubes with an equivalent weight of water, and autoclaved at 600 °C and 3.20 kbar for 24-32 hr. Products were recovered by filtration, using Gelman 0.22 μm membrane filters. Portions of the hydrothermally synthesized YAG:Tb samples were annealed at 1200 °C and 1450 °C.

$NaY(WO_4)_2$:Tb compositions were synthesized at high temperature by combining stoichiometric quantities of Y_2O_3, $Tb(NO_3)_3{\bullet}5H_2O$, Na_2WO_4, and WO_3, then heating the mixture to 1000 °C. These phases were also prepared hydrothermally by reacting solutions of Y and Tb nitrates with Na_2WO_4 in water at 200 °C and 15 bar. Portions of the hydrothermally synthesized samples were annealed at 900 °C.

2.2. Characterization

Particle sizes were determined by scanning electron microscopy (SEM) and by sedimentation. Samples for SEM were sputtered with Au and analyzed using an AMRAY model 1645 electron microscope. Particle size determination by sedimentation was accomplished by suspending the powder samples in a sodium pyrophosphate solution, sonicating, and measuring settling rates using a Horiba CAPA-700 particle size analyzer. Powder X-ray diffraction (XRD) data were collected using a Scintag Pad V diffractometer. Crystallite sizes were calculated from the widths of the (420) and (101) diffraction peaks of $Y_3Al_5O_{12}$ and $NaY(WO_4)_2$, respectively, using a LaB_6 standard.

Photoluminescence (PL) and photoluminescence excitation (PLE) spectra were collected from the phosphor powders at room temperature using a computer-driven monochromator system. The output from a tunable excitation source (Xe arc lamp coupled to a 0.15 m monochromator equipped with order-sorting filters) was imaged to a spot size of approximately 0.8 x 5 mm at the sample with a spectral bandpass of 4 nm. PL spectra were acquired at an excitation wavelength of 254 nm. During PLE data collection, the 0.15 m (excitation source) monochromator slits were set to produce a 2 nm spectral bandpass. The resulting luminescence was analyzed using a 0.275 m single-pass monochromator equipped with a photomultiplier tube detector. The collection monochromator resolution was typically 0.5 nm for the survey spectra obtained. A lock-in amplifier and mechanical chopper provided phase-sensitive detection of the optical signal. Fluctuations in the lamp output were mediated using a beam splitter in the excitation beam path to divert a portion of the incident beam to a separate Si photodiode. The output of this detector was used to normalize the luminescence signal during data acquisition.

Cathodoluminescence data were collected in a vacuum chamber evacuated to $4x10^{-6}$ Pa. A hot filament, low-energy electron gun (Kimball Physics) was the source of the beam, which was steered and focused using external Helmholtz coils. The powder samples were packed approximately 1 mm deep into stainless steel cups. During analysis the sample cup was maintained at a potential of +30 V with respect to ground. The electron beam was focused onto a spot 4.5 mm in diameter. Light emitted from the sample was transmitted to a spectroradiometer (Oriel), which measured radiant energy as a function of wavelength.

3. RESULTS

3.1. Synthesis

In the course of hydrothermal processing the amorphous YAG precursor gel crystallizes to form regular grains with a garnet habit. Grains consist of

Figure 1a: Hydrothermally synthesized YAG:Tb powders (scale bar = 0.375 μm).

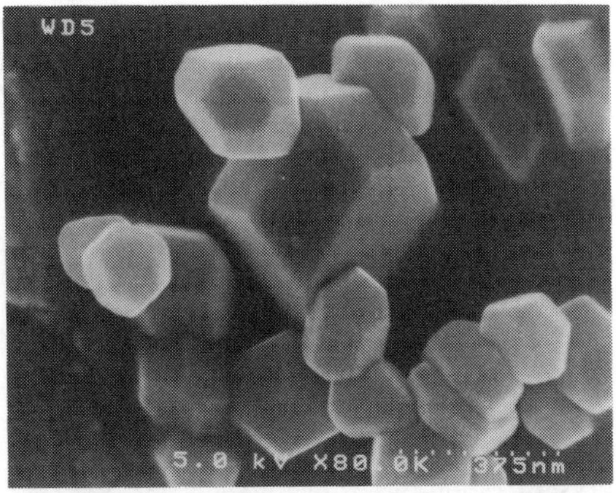

Figure 1b: Hydrothermally synthesized YAG:Tb, annealed at 1450 °C (scale bar = 1 μm).

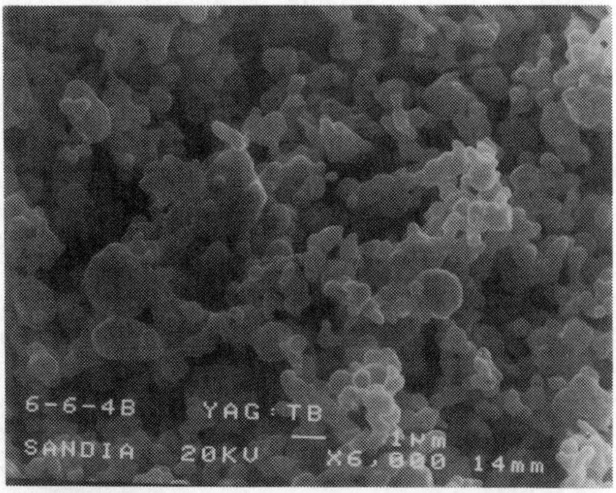

Figure 1c: Hydrothermally synthesized YAG:Tb grains from over-nucleated precursor gel (scale bar = 1 μm).

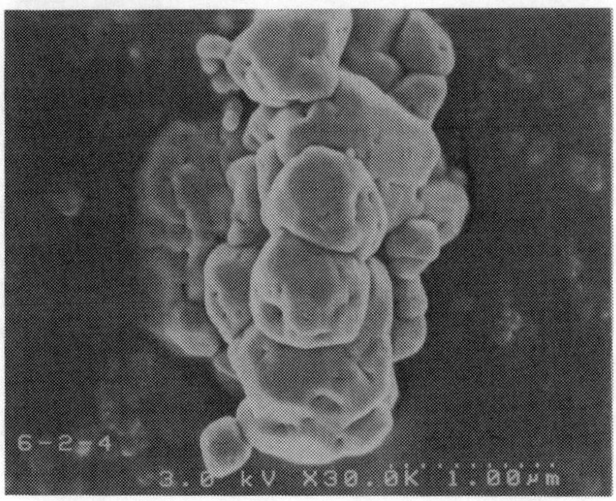

well-formed YAG:Tb crystallites with particle diameters between 0.2 and 2 μm, with the largest mass fraction at 0.5 μm (Figure 1a). These crystallites showed little tendency to agglomerate, and could be readily suspended in water through sonication. Minor sintering of smaller crystallites was evident at 1200 °C, while the powders annealed at 1450° were heavily sintered (Figure 1b).

It is pointed out in ref. 10 that if the YAG:Tb gel is heated beyond ca. 600 °C, it will become excessively nucleated, though the powder will remain X-ray amorphous until it is heated to ca. 900 °C. To determine the effect of crystallization of the precursor gel on the product YAG:Tb grains, a YAG:Tb precursor gel was heated to 700 °C, yielding a powder that is X-ray amorphous. This material was autoclaved in water in the same fashion as the amorphous gel. The resulting YAG:Tb powder consisted of rough polycrystalline grains with a crystallite size of 56 nm (Figure 1c). This result is likely due to increased nucleation density in the precursor, and possibly to decreased solubility of crystalline YAG in water versus the YAG precursor gel.

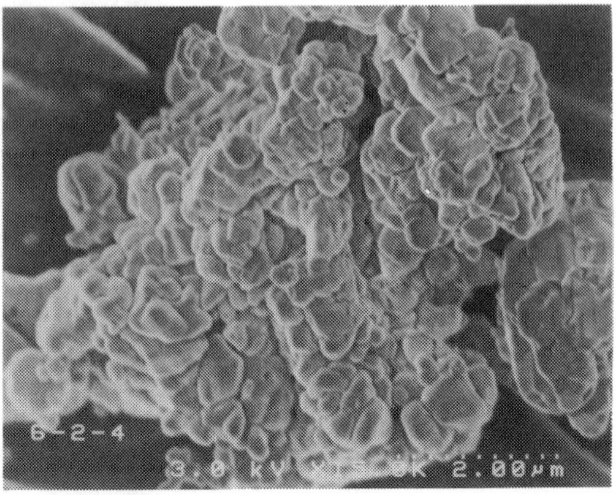

Figure 2a: YAG:Tb grain synthesized via solid state reaction at 1200 °C (scale bar = 2 μm).

High-temperature synthesis of YAG:Tb via the solid state yields irregular grains with a typical diameter of 5 μm. These grains give the appearance of sintered microcrystallites in SEM views (Figure 2a). The effective crystallite size of YAG:Tb prepared at 1200 °C is 53 nm. Annealing at 1450 °C causes further sintering to yield crystallites with an average diameter of 320 nm (Figure 2b).

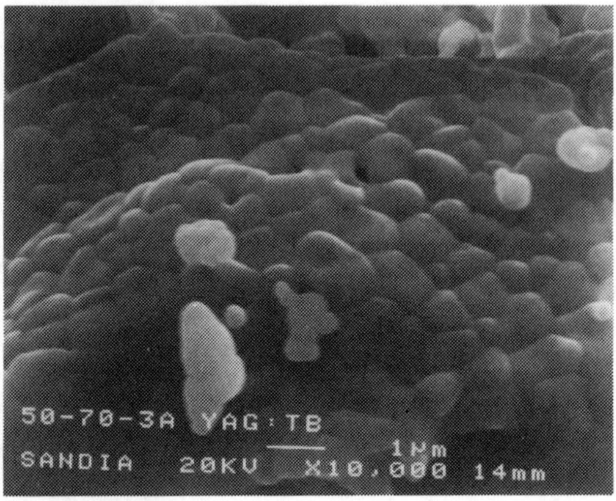

Figure 2b: Solid-state YAG:Tb, annealed at 1450 °C (scale bar = 1 μm).

Hydrothermal synthesis of $NaY(WO_4)_2$:Tb yields particles consisting of intergrown plates (Figure 3a). The effective crystallite size in these particles is approximately 800 nm. Annealing at 900 °C causes intraparticulate sintering to yield grains several microns across (Figure 3b). The grains thus obtained have essentially the same size and morphology as those prepared via high-temperature solid state reaction.

Figure 3a: $NaY(WO_4)_2$:Tb grains, synthesized hydrothermally at 200 °C (scale bar = 10 µm).

3.2. Luminescence

The luminescence center in a phosphor may undergo excitation and emission through several different mechanisms. These are generalized here as direct activation, in which incident energy is absorbed and reradiated by localized species (such as an activator ion, sensitizer, or charge-transfer band) and indirect activation, in which incident energy creates an excitation in the host lattice that is subsequently transferred to the luminescence center. A phosphor that is capable of indirect activation can also be directly activated if the excitation energy is detuned from the host lattice absorption and is made coincident with allowed transitions in the activator ion. Using optical excitation (PL), both direct or indirect mechanisms can be probed by varying the excitation energy. For example, Y_2O_3:Eu can be directly activated using 300 nm light through electronic transitions in the Eu^{3+} dopant. Conversely, the phosphor is indirectly activated at 200 nm through host lattice absorption.[17]

Figure 3b: Hydrothermally synthesized $NaY(WO_4)_2$:Tb after annealing at 900 °C (scale bar = 10 µm).

While both types of excitation may be accessed using PL, the cathodoluminescence process deposits energy only near the surface of a phosphor grain through inelastic scattering with lattice ions to produce secondary electrons, which then generate electron-hole pairs in the lattice. The excitation (in the form of electron-hole pairs) is then propagated through the host lattice to dopant ions within the grain interior to indirectly achieve activator excitation and radiative emission. Relative changes in the mean free path of this indirect excitation as it interacts with activator ions and microstructural features (such as defects, impurities, or grain boundaries) impact the luminescence efficiency of the phosphor. If recombination occurs nearly exclusively at an activator site, the average migration distance of the excitation will vary inversely with activator concentration. At lower activator concentrations (or smaller crystal sizes), migration distance and thus activation volume will be limited by crystal size, and at higher concentrations it will be limited by the density of activators. As activator concentration increases to the point where the density of activator ions begins to limit activation volume, PL and CL intensities becomes less sensitive to activator concentration.[18] This work examines the activation mechanisms for YAG:Tb and $NaY(WO_4)_2$:Tb phosphor powders with varied dopant concentrations through both PL and CL efficiency measurements.

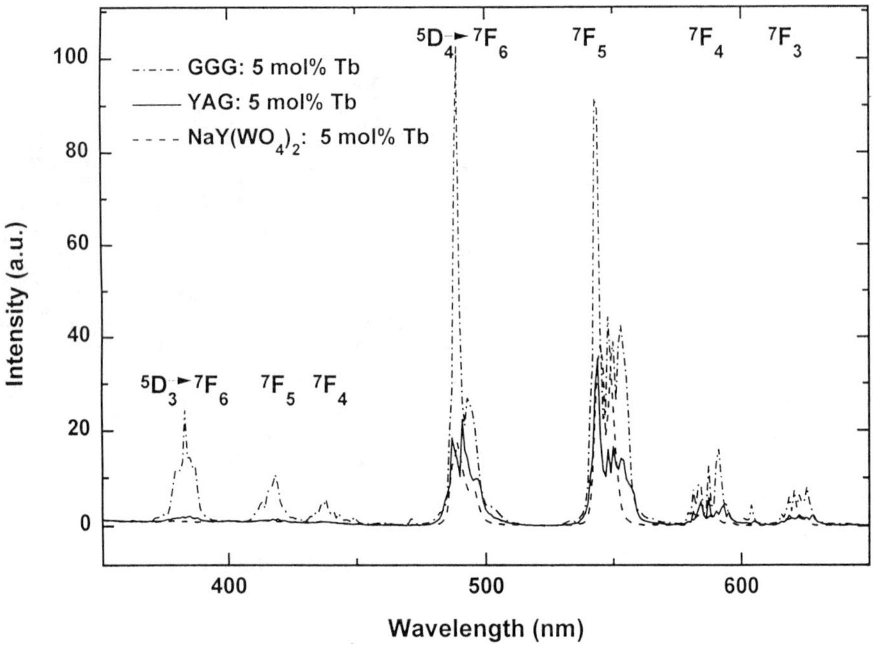

Figure 4: Photoluminescence spectra of $Y_3Al_5O_{12}$:Tb, $Gd_3Ga_5O_{12}$:Tb, and $NaY(WO_4)_2$:Tb, acquired using a 254 nm excitation source.

3.2.1. Photoluminescence: PL spectra (Figure 4) show that over the UV to visible wavelength range, the integrated luminescence intensities for $Y_{2.85}Tb_{0.15}Al_5O_{12}$, $Gd_{2.85}Tb_{0.15}Ga_5O_{12}$, and $NaY_{0.95}Tb_{0.05}(WO_4)_2$ increase in the order $NaY(WO_4)_2$ < YAG < GGG for excitation at 254 nm. Figure 5 shows the dependence of PL intensity on Tb concentration measured for the YAG and $NaY(WO_4)_2$ hosts. While the dependence of PL vs. concentration is steeper in $NaY(WO_4)_2$:Tb than in YAG:Tb, both hosts appear to reach similar intensity values at high dopant concentrations. A maximum integrated intensity was observed in the $NaY(WO_4)_2$ sample when 50% of the Y was replaced with Tb. This contrasts with the general increase in intensity observed in the YAG host over this concentration range. The $NaY(WO_4)_2$:Tb data appear linear (within error limits) until they exhibit a decrease for the sample with the highest Tb concentration. This is presumably due to concentration quenching effects. The linearity is indicative of the direct activation mechanism used in the experiment. No significant increase in integrated PL intensity was observed after the annealing treatments. This suggests that there was no significant change in activator local environment that led to a variation in transition oscillator strength.

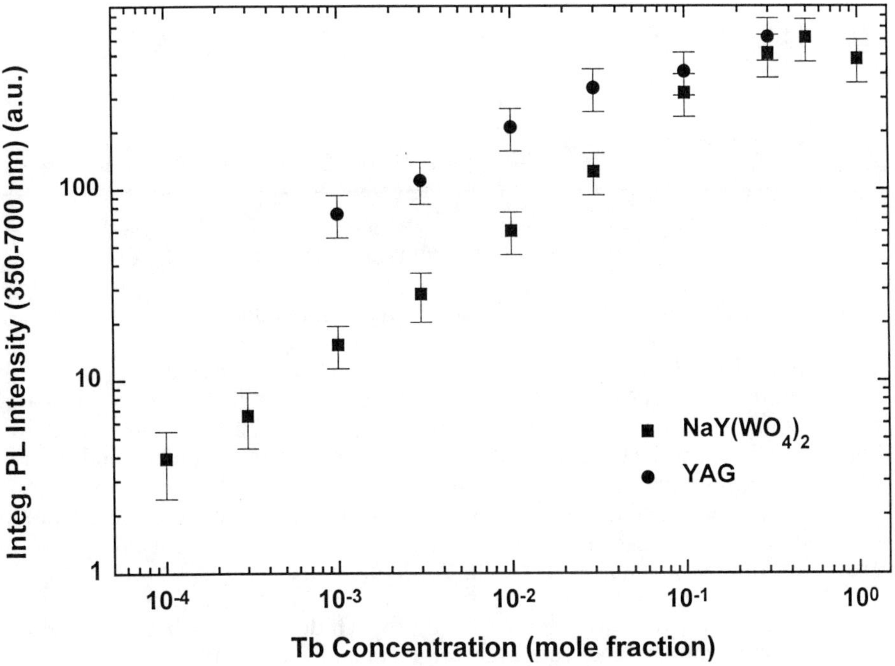

Figure 5: PL intensity vs. Tb concentration in $NaY(WO_4)_2$:Tb and $Y_3Al_5O_{12}$:Tb.

These initial conclusions regarding the excitation mechanisms in these phosphor samples were tested using photoluminescence excitation (PLE) measurements. Luminescence at 550 nm was monitored while scanning the excitation source. Figure 6 shows that a maximum in 550 nm output was observed with excitation near 275 nm for $NaY(WO_4)_2$:Tb (similar results were obtained for YAG:Tb).

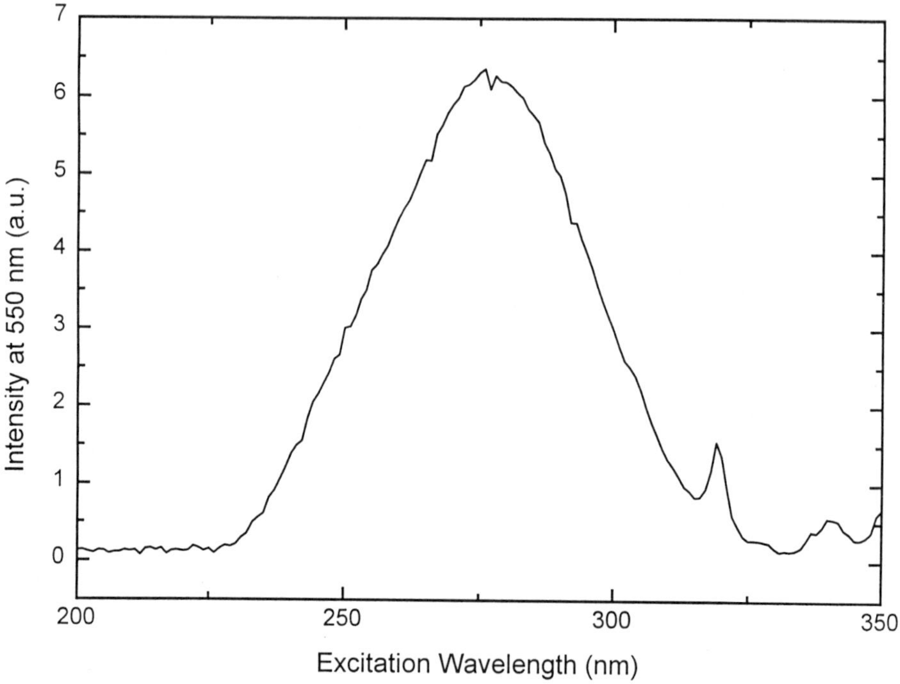

Figure 6: PL excitation spectrum of $NaY(WO_4)_2$:Tb, monitored at 550 nm.

This broad PLE band results from 4f to 5d interband transitions within the rare-earth dopant ion. Smaller features at longer wavelengths correspond to 4f-4f absorption processes. Thus, only direct excitation of the Tb^{3+} ions is attained optically. Diffuse reflectance from the samples indicate absorption edge onset in the 220 to 250 nm range.[19] While no features are observed at shorter wavelengths in the PLE spectra, an estimation of the relative influence of host lattice absorption on the indirect excitation of Tb ions is not made here due to the significant decrease in excitation intensity available from the Xe lamp source at this wavelength range. Given the PLE data, however, it is apparent that the 254 nm excitation energy used in the PL experiment is directly activating the Tb dopant, which is in agreement with the concentration dependence of the integrated PL intensity.

3.2.2. Cathodoluminescence: CL efficiency data measured as a function of electron voltage at a constant power of 1.5×10^{-5} W for YAG:Tb and GGG:Tb powders are presented in Figure 7. CL efficiency increases with voltage at constant power, approximately doubling between 150 and 1000 V. The threshold voltage for all phosphors was between 100 and 150V. The efficiencies of YAG:Tb prepared hydrothermally and via the solid state were similar to each other at all voltages, while those of GGG:Tb were greater by approximately 0.5 lm/W. The dead voltage and steep slope of the efficiency vs. voltage curve are most likely due to surface bound electrons: at higher voltages, luminance is higher at constant power due to increased penetration depth of the primary electron beam.

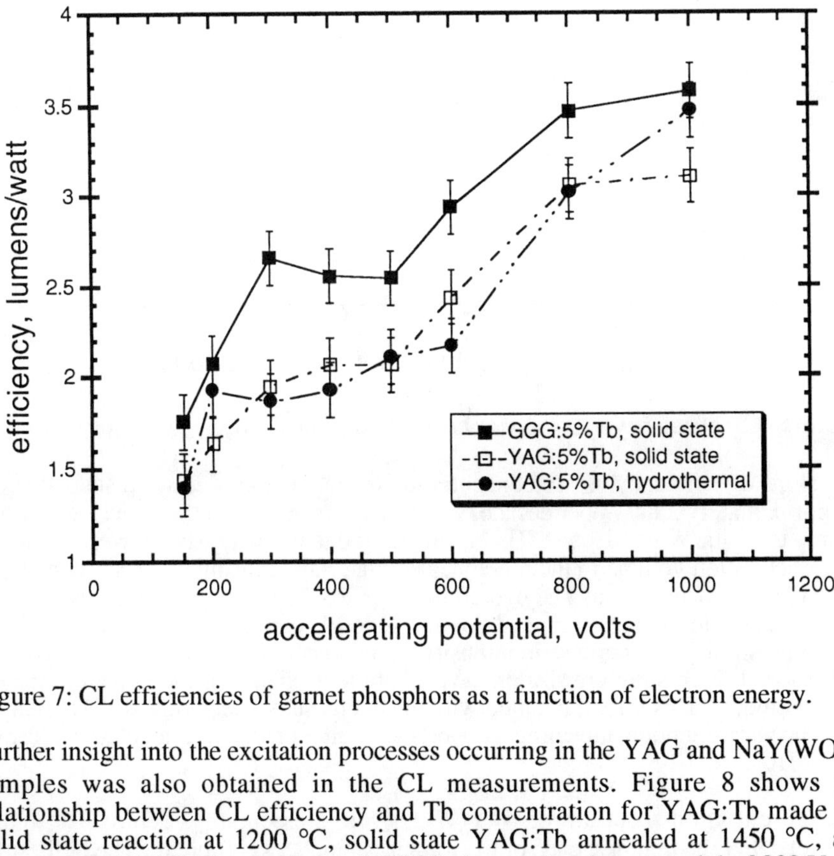

Figure 7: CL efficiencies of garnet phosphors as a function of electron energy.

Further insight into the excitation processes occurring in the YAG and $NaY(WO_4)_2$ samples was also obtained in the CL measurements. Figure 8 shows the relationship between CL efficiency and Tb concentration for YAG:Tb made via solid state reaction at 1200 °C, solid state YAG:Tb annealed at 1450 °C, and hydrothermal YAG:Tb annealed at 1200 °C. An accelerating potential of 800 V and beam power of $1.5x10^{-5}$ W was used for all data shown. The efficiencies of the annealed solid state and hydrothermal samples are similar at all concentrations, except that the samples made via the solid state are more susceptible to quenching at the highest dopant levels. Maximum efficiency is found at dopant concentrations between 5 and 10 mole percent in both solid state and hydrothermal YAG:Tb.

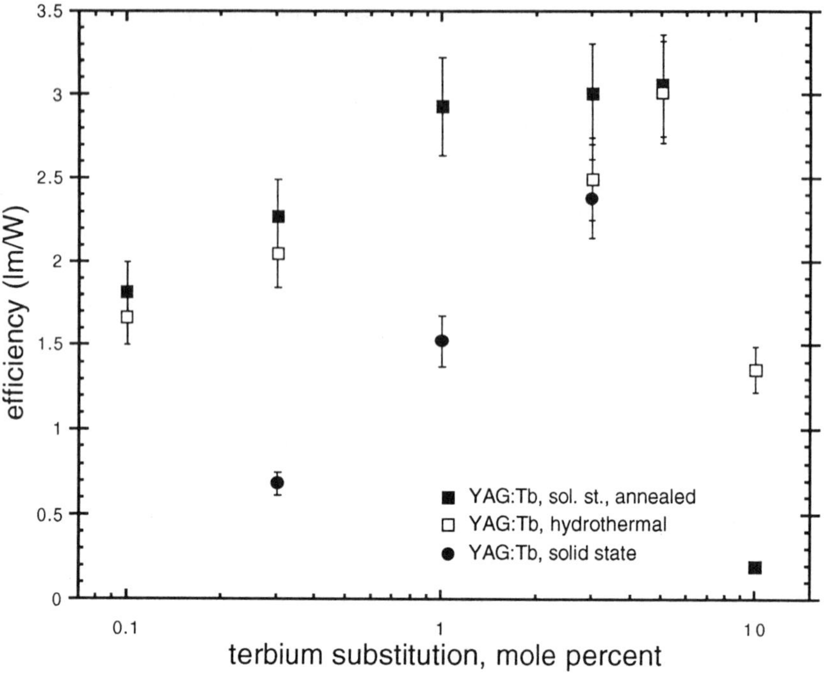

Figure 8: CL efficiency vs. activator concentration in $Y_3Al_5O_{12}$:Tb phosphors.

The slope of efficiency vs. Tb concentration for the unannealed solid state samples is approximately 2 lm/W per decade, while this dependence in the annealed samples is only ca. 1 lm/W per decade. This is probably due to the crystal size effect, which limits activation volume in the pre-annealed materials, causing increased sensitivity to activator concentration. The overall increase in CL intensity after annealing over the concentration range examined contrasts the PL data from the same samples in which no significant increase in intensity is observed. This indicates the increased influence of microstructure size on the excitation efficiency of the dopant in the CL measurement. This effect, coupled with the decreased dependence of CL intensity on concentration upon annealing, supports an indirect activation mechanism for the YAG:Tb.

The CL emission spectrum of $NaY(WO_4)_2$:Tb is essentially identical to that of YAG:Tb, but the luminescence of the former is much less efficient. The highest CL intensities observed for the $NaY(WO_4)_2$:Tb compositions were approximately 1% of that found for YAG:Tb. This again contrasts the PL studies, which show a much smaller difference in integrated luminescence intensity between the two host materials (particularly at the higher concentrations). The difference between the CL and PL trends reflects the predominant excitation mechanisms at work in these studies. The significantly higher CL intensities for the YAG host indicate that it is indirectly activated by the electron beam much more efficiently than $NaY(WO_4)_2$, which appears to rely on direct activation of the Tb dopant in order to luminesce.

4. DISCUSSION

YAG:Tb and GGG:Tb are efficient CL phosphors at low voltages (< 1000 V). The PL and CL intensities from GGG:Tb were both greater than those from YAG:Tb; CL efficiencies as high as 3.5 lm/W were observed with GGG:Tb. By depositing these phosphors on conducting screens, it may be possible to further improve their performance, particularly at lower voltages (< 500 V). While the efficiency values reported here are still lower than those of the best green-emitting sulfide and oxysulfide phosphors, their excellent chromaticities and resistance to degradation and power saturation warrant consideration of garnet phosphors for use in field-emitter flat panel display screens.

Synthesizing YAG phosphors hydrothermally does not materially affect CL efficiency, so any reduction in surface recombination centers from low-temperature processing are probably not significant. It is more likely that surface-bound electrons are the principal cause of efficiency loss at lower voltages. Compared with conventional high-temperature processing, though, hydrothermal synthesis yields smaller crystallites of YAG with improved particle morphology. This would benefit projection CRT screens, which currently employ YAG:Tb as green-emitting phosphors.

The mechanism of cathodoluminescence requires transfer of absorbed electron energy from the host lattice to the activator, that is, a CL phosphor must be capable of being indirectly activated. While the PL intensities of Tb emission from YAG:Tb and $NaY(WO_4)_2$:Tb are similar, the CL efficiency of $NaY(WO_4)_2$:Tb is two orders of magnitude lower than that of YAG:Tb. These results strongly suggest that this phosphor can only be directly activated. While the low CL emission intensities from $NaY(WO_4)_2$:Tb may proscribe its use as a CL phosphor, its high PL efficiency may make it usable with UV-emitting phosphors as a wavelength shifter for lamps and displays.

5. ACKNOWLEDGMENTS

This work was supported by the U. S. Department of Energy under contract no. DE-AC04-94AL85000, and benefited from the assistance of R. J. Walko, R. Mays, G. R. Schuster, L. E. Shea, G. L. Zender, and D. L. Lamppa.

6. REFERENCES

1. *a.* I. Brodie and C. A. Spindt, "Vacuum Microelectronics" pp. 1-106 in Advances in Electronics and Electron Physics, Vol. 83, P. W. Hawkes, Ed., Academic Press: San Diego, 1992. *b.* C. A. Spindt, C. E. Holland, A. Rosengreen, and I. Brodie, "Field-Emitter Arrays for Vacuum Microelectronics",

IEEE Trans. Electron Dev. 38(10), 2355-2363 (1991). *c.* C. A. Spindt, C. E. Holland, I. Brodie, J. B. Mooney, and E. R. Westerberg, "Field-Emitter Arrays Applied to Vacuum Fluorescent Display", *IEEE Trans. Electron Dev.* 36(1), 225-228 (1989).

2. S. Yang, F. Zhang, C. Stoffers, S. M. Jacobsen, C. J. Summers, P. N. Yocom, S. McClelland, "Characterization of Potential Low-Voltage Phosphors for Field Emission Devices", *Proc. SPIE - Int. Soc. Opt. Eng.* 2408, 194-199 (1995).

3. S. Itoh, T. Kimizuka, and T. Tonegawa, "Degradation Mechanism for Low Voltage Cathodoluminescence of Sulfide Phosphors", *J. Electrochem. Soc.* 136(6), 1819-1823 (1989).

4. S. Itoh, T. Niiyama, and M. Yokoyama, "Influence of Gases on the Field Emission", *J. Vac. Sci. Technol.* B11, 647-650 (1992).

5. *a.* L. Ozawa and S. Hayakawa, "A Study on Cathodoluminescence Intensities in the Range Below 3 kv", *J. Lumin.* 28, 327-335 (1983).

6. S. Itoh, T. Tonegawa, T. L. Pykosz, K. Morimoto, and H. Kukimoto, "Influence of Grinding and Baking Processes on the Luminescent Properties of Zinc Cadmium Sulfide Phosphors for Vacuum Fluorescent Displays", *J. Electrochem. Soc.* 134 (12), 3178-3181 (1987).

7. Efficiency data: a. K. Narita, A. Kagami, and Y. Mimura, "Behavior of Phosphors under Low Voltage Cathode Ray Excitation", *J. Electrochem. Soc.* 127 (8), 1794-1798 (1980). b. A. O. Dmitrienko and S. L. Shmakov, "The Efficiency of Low-Voltage Cathodoluminescence of Modified Zinc Oxide Crystallophosphors", *Inorganic Materials* 30 (4), 534-535 (1994). Threshold voltage: R. E. Shrader and S. F. Kaisel, "Excitation of Zinc Oxide Phosphors by Low-Energy Electrons", *J. Opt. Soc. Am.* 44 (2), 135-137 (1954).

8. R. Raue, A. T. Vink, and T. Welker, "Phosphor Screens in Cathode-Ray Tubes for Projection Television", *Philips Tech. Rev.* 44 (11/12), 335-347 (1989).

9. $NaY(WO_4)_2$ as a laser host: H. Yizong, G. Wang, X. Huang, Z. Luo, "The Emission and Structure of a New Laser Crystal Nd^{3+}:$NaY(WO_4)_2$", *Cryst. Res. Technol.* 29(2), 267-271 (1994). $NaY(WO_4)_2$ as an anti-Stokes phosphor host: *a.* F. Auzel and D. Pencile, "Comparison and Efficiency of Materials for Summation of Photons Assisted by Energy Transfer", *J. Lumin.* 8(1), 32-43 (1973). *b.* B. M. Tissue, L. Lu, L. Ma, W. Jia, M. L. Norton, W. M. Yen, "Laser-heated Pedestal Growth of Laser and IR-Upconverting Materials", *J. Cryst. Growth* 109, 323-328 (1991).

10. T. Takamori and L. D. David, "Controlled Nucleation for Hydrothermal Growth of Yttrium-Aluminum Garnet Powders", *Am. Ceram. Soc. Bull.* 65 (9), 1282-1286 (1986).

11. E. D. Kolb and R. A. Laudise, "Phase Equilibria of $Y_3Al_5O_{12}$, Hydrothermal Growth of $Gd_3Ga_5O_{12}$, and Hydrothermal Epitaxy of Magnetic Garnets", *J. Cryst. Growth* 29, 29-39 (1975).

12. "Procédé de préparation d'oxydes métalliques complexes", French Patent 2,007,316 (1970, to Swiss Aluminium Ltd.)

13. T. Takahashi, "Columnar crystalline zinc silicate phosphor prepared by hydrothermal reaction", Japanese Patent 63,196,683 (1988).

14. M. N. Viswanathiah, J. A. K. Tareen, and T. R. Narayanan Kutty, "Hydrothermal Phase Equilibria Studies in Ln_2O_3-H_2O Systems and Synthesis of Cubic Lanthanide Oxides", *Mater. Res. Bull.* 15(7), 855-859 (1980).

15. *a.* R. C. Ropp and B. Carroll, "Yttrium Phosphate-Yttrium Vanadate Solid Solutions and Vegard's Law", *Inorg. Chem.* 14(9), 2199-2202 (1975). *b.* O. Yamaguchi, Y. Mukaida, H. Shigeta, H. Takemura, and M. Yamashita, "Preparation of Alkoxy-Derived Yttrium Vanadate", *J. Electrochem. Soc.* 136(5), 1557-1560 (1989). *c.* H. J. Kaunders, "Europium-doped Yttrium Vanadate(V) Phosphor Preparation by Precipitation", U. S. Patent 3,479,296 (1969, to General Electric). *d.* M. L. F. Phillips, "Low-Voltage Cathodoluminescence of Europium-Activated Yttrium Orthovanadate", *Proc. SPIE - Int. Soc. Opt. Eng.* 2408, 200-206 (1995).

16. R. C. Ropp, The Chemistry of Artificial Lighting Devices, pp. 502-504, Elsevier: Amsterdam, 1993.

17. L. Ozawa, Cathodoluminescence, pp. 20-21, VCH Publishers: New York, 1990.

18. *a.* L. Ozawa and H. N. Hersh, "Random-Walk Model of Energy Transfer in Cathodoluminescence", *Phys. Rev. Lett.* 36(12), 683-686 (1976). *b.* L. Ozawa, "An Experimental Proof of Radiative Recombination Mechanisms of Electron-Hole Pairs at Activator Ions", *J. Electrochem. Soc.* 128(1), 140-145 (1981).

19. M. L. F. Phillips and B. G. Potter, unpublished results.

EFFECT OF PROCESSING PARAMETERS ON LUMINESCENCE OF COMBUSTION-SYNTHESIZED, RARE EARTH-ACTIVATED $Y_3Al_5O_{12}$ PHOSPHORS

L. E. Shea, J. McKittrick, and O. A. Lopez
Department of Applied Mechanics and Engineering Sciences and Materials Science Program, University of California, San Diego, La Jolla, California 92093-0411

Esther Sluzky
Hughes Aircraft Company, Carlsbad, California 92008

ABSTRACT

A novel ceramic synthesis technique, combustion synthesis, was explored to produce $Y_3Al_5O_{12}$ (YAG) phosphors with improved physical and luminescent properties. The rare-earth ions Eu^{3+}, Tb^{3+}, and Tm^{3+} were used as activators for red, green, and blue emission respectively. The combustion synthesis technique involves the reaction of metal nitrates (oxidizers) and an organic fuel at 500°C. Resulting powders are well-crystallized, with small particle size. Luminous intensity was observed using photoluminescence measurements. The effects of processing parameters such as type of fuel, ignition temperature, and and activator concentration were studied. Particle sizes were found to range from 0.4 to 3 μm, depending upon the processing parameters.

INTRODUCTION

Rare-earth activated phosphors find application in fluorescent lamps, cathode-ray tubes (CRTs), heads-up displays, and flying spot scanners. High resolution, short decay time, and high brightness are major requirements for information display devices. To achieve high resolution, uniform, small particle size (< 3μm), phosphors are desirable [1,2]. Conventional phosphor synthesis techniques cannot always achieve these requirements, hence, the development of both novel and improved phosphor synthesis techniques is a current research interest. The objective of this work was to produce rare-earth activated yttrium aluminum garnet (YAG) phosphors with high brightness and small particle size, more efficiently using a novel combustion synthesis technique.

This technique appeared promising for phosphor synthesis because of its success in producing fine particle size, multicomponent, homogeneous oxide ceramics at considerably lower temperatures and with reduced processing time [3-8]. The combustion process involves the exothermic reaction of an oxidizer such as metal nitrates, ammonium nitrate, and ammonium perchlorate [8], and an organic fuel, typically urea (CH_4N_2O), carbohydrazide (CH_6N_4O), and glycine ($C_2H_5NO_2$). The reaction is initiated at temperatures much lower than the phase transition of the target material. In a typical reaction, the precursor mixture decomposes, dehydrates, and ruptures into a flame in less than 5 minutes. The product is a voluminous, foamy powder which occupies the volume of the reaction vessel. The chemical energy released from the decomposition of the metal nitrates and fuel can rapidly heat the system to high temperatures (>1600°C) and sustain these high temperatures without an external heat source.

EXPERIMENTAL PROCEDURE

Phosphors with the general formula $(Y_{1-x}Ln_x)_3Al_5O_{12}$ (YAG:Ln) where Ln = Eu^{3+}, Tb^{3+}, or Tm^{3+}, with x = 0.05 for Eu^{3+}, x = 0.075 for Tb^{3+}, and x = 0.2 to 5.0 for Tm^{3+}, were prepared by combustion synthesis. The starting materials $Y(NO_3)_3 \cdot 6H_2O$ (Alfa REacton 99.99%), $Al(NO_3)_3$ (Alfa Aesar 99.999%), $Ln(NO_3)_3 \cdot 6H_2O$ (Alfa Puratronic 99.999%) were thoroughly mixed in a mortar and pestle with the appropriate amount of fuel. To prevent water uptake, mixing was done under nitrogen in a vacuum glove box. The starting materials were placed in a 300 mL Pyrex dish with 0.667 mL H_2O/g of nitrate, and introduced into a muffle furnace at 500°C. After the combustion reaction, which occurred in 3-5 min., the resultant powders were re-mixed, placed in 50 mL alumina crucibles and heat treated for 1 h. at temperatures up to 1600°C.

Powders were characterized by x-ray diffraction, BET surface area analysis, and carbon-nitrogen elemental analysis. Photoluminescence emission spectra were obtained using a Hg lamp.

RESULTS AND DISCUSSION

For combustion reactions using carbohydrazide and urea, the YAG phase crystallized directly from the reacted precursor without forming any intermediate phases. The XRD pattern of as-synthesized YAG:Ln powders shows phase pure YAG as shown in Figure 1. No other phases could be detected.

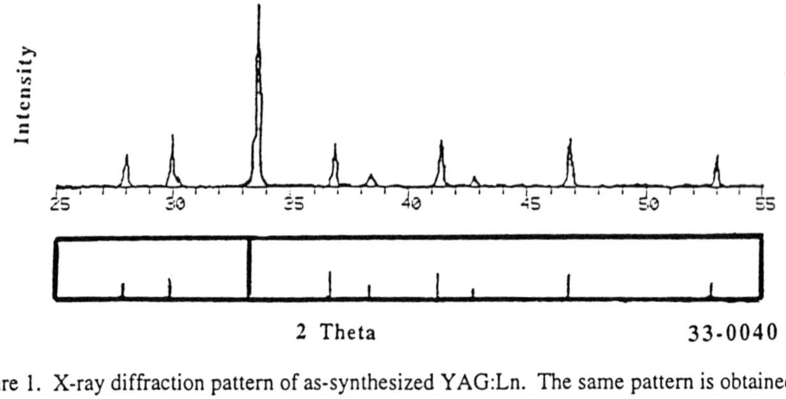

Figure 1. X-ray diffraction pattern of as-synthesized YAG:Ln. The same pattern is obtained for Ln=Eu, Tb, Tm.

When glycine fuel was used, an amorphous black ash was produced in the as-synthesized condition. The YAG structure did not crystallize from the ash until 1000°C. The light output of YAG phosphors made with urea, carbohydrazide and glycine was compared. The brightest phosphors were obtained with carbohydrazide. Particle size from BET measurements were 3.3, 0.9, and 0.45 μm for phosphors made with urea, glycine, and carbohydrazide respectively.

Figure 2 shows the emission spectra of as-synthesized YAG:Tm^{3+} made with carbohydrazide. YAG phosphors with low Tm^{3+} concentration, exhibit a red emission band (650-700 nm) and weak blue emission (450-500 nm). As more Tm^{3+} ions are incorporated, (up to 5.0 m/o), the red emission decreases while the blue emission increases, due to more efficient energy transfer between the Tm^{3+} ions [9]. For YAG:Tm^{3+} to be used as a blue phosphor, a high Tm^{3+} concentration would be needed to intensify the blue peak and quench the red.

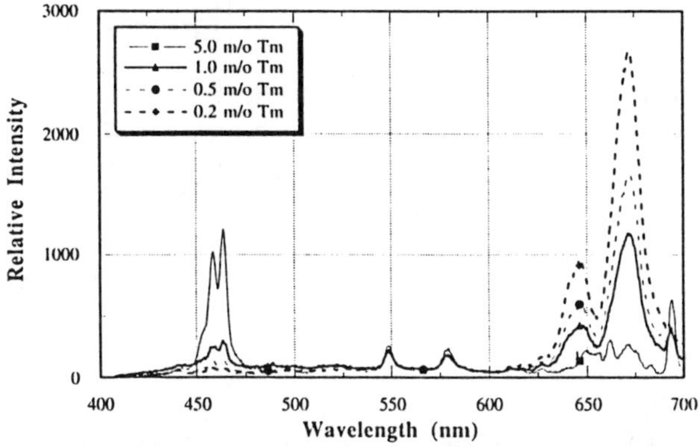

Figure 2. Effect of Tm^{3+} concentration on YAG:Tm^{3+} emission spectra under 365 nm excitation

Heat treatments to 1600°C were necessary to maximize the brightness, as shown in Figure 3 for YAG:Eu^{3+} phosphors made with carbohydrazide. Carbon and nitrogen impurities from the combustion reaction were investigated as a possible cause of the lower brightness of as-synthesized powders. High-purity is a crucial requirement for phosphors, since impurities can absorb the excitation energy, and quench the luminescence process. Table I lists the weight percentages of carbon and nitrogen for as-synthesized and heat treated YAG:Ln powders made with carbohydrazide, glycine, and urea.

Table I. Elemental Analysis of YAG phosphors

	Carbohydrazide		Glycine		Urea	
Heat Treatment	Weight % C	Weight % N	Weight % C	Weight % N	Weight % C	Weight % N
As-synthesized	0.08	0.12	7.02	2.63	0.07	0.10
1000°C	0.03	0.05	0.14	0.09	0.04	0.08
1300°C	0.03	0.04	0.08	0.03	0.04	0.01
1600°C	0.03	0.01	0.02	0.01	0.04	0.01

As expected, since the body color of powders made with glycine were black, there is quite a high carbon content. However, all materials have only residual carbon and nitrogen impurities after heat treatment to 1600°C.

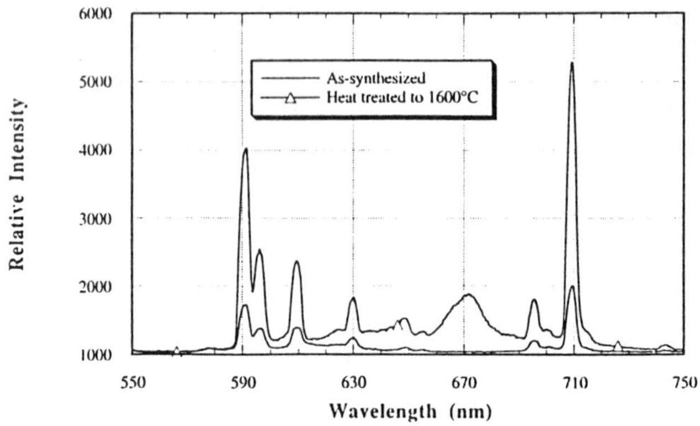

Figure 3. Emission spectra of as-synthesized and heat treated YAG:Eu^{3+} (435 nm excitation)

CONCLUSIONS

A combustion synthesis technique has been used to produce single phase, well-crystallized, rare earth-activated YAG phopshors with high brightness and small particle size. Post reaction heat treatments are required to maximize brightness. Carbon content may affect the light output of as-synthesized phosphors by absorbing a portion of the excitation energy, but this is substantially reduced after annealing to 1600°C.

ACKNOWLEDGMENTS

Financial support of this work from the California Micro Program and Hughes Aircraft Company, Carlsbad, CA. is gratefully acknowledged. Thanks is given to Mike Sailor of the UCSD chemistry department for use of his laboratory, and to J. J. Kingsley for helpful discussion.

REFERENCES

1. N. A. Diakides, Proc. of the Soc. of Photo-Optical Instrumentation Engineers 17th Annual Technical Meeting, **42**, 83 (1973).
2. R. Raue, A. T. Vink, and T. Welker, *Philips Tech Rev.*, **44** [11/12], 325 (1989).
3. L. A. Chick, L. R. Pederson, G. D. Maupin, J. L. Bates, L. E. Thomas, and G. J. Exarhos, *Materials Letters*, **10** [1,2], 6 (1990).
4. L. R. Pederson, G. D. Maupin, W. J. Weber, D. J. McCready, and R. W. Stephens, *Materials Letters*, **10** [9], 437 (1991).
5. N. J. Hess, G. D. Maupin, L. A. Chick, D. S. Sunberg, D. E. McCreedy, and T. R. Armstrong, Submitted to J. Mater. Sci., 1993.
6. J. J. Kingsley and K. C. Patil, *Materials Letters*, **6** [11,12], 427 (1988).
7. J. J. Kingsley, N. Manickam, and K. C. Patil, *Bull. Mater. Sci.*, **13** [3], 179 (1990).
8. Y. Zhang and G. C. Stangle, *J. Mater. Res.*, **9** [8], 1997 (1994).
9. S. Guy, M. F. Joubert, and B. Jacquier, *Phys. Stat. Sol. B.*, **183** [K33] (1993).

RELATIONSHIP BETWEEN MICROSTRUCTURE AND EFFICIENCY OF LITHIUM SILICATE SCINTILLATING GLASSES: THE EFFECT OF ALKALINE EARTHS

M. BLISS, R. A. CRAIG, D. S. SUNBERG, and M. J. WEBER
Pacific Northwest National Laboratory, Box 999, Richland, WA 99352

ABSTRACT

Lithium silicate glasses containing Ce^{3+} are known to be scintillators. Glasses in this family in which the Li is enriched (^{6}Li) are used as neutron detectors. The addition of Mg to this glass is known to increase the scintillation efficiency. We have found that substituting other alkaline earths results in a monotonic decrease of the scintillation efficiency with increasing atomic number. The total variation in scintillation efficiency from Mg to Ba is nearly a factor of 3. Prior experiments with this glass family show small differences in Raman and fluorescence spectra; evidence from thermoluminescence experiments indicates that the scintillation efficiency is most strongly correlated with structural effects in the neighborhood of the Ce^{3+} activator ion. The results of low-temperature studies of fluorescence and thermoluminescence of these glasses will be reported.

INTRODUCTION

Cerium-activated, lithium-silicate glasses constitute an important and interesting system of scintillators.[1] Small changes in composition can have dramatic changes in scintillation efficiency. For example, adding Mg to these glasses increases scintillation efficiency. Bliss et al.[2] (hereinafter [I]) have shown that substituting a different alkaline earth for the Mg results in a monotonic decrease in scintillation efficiency with atomic number. Changing the network modifier from one alkaline earth to another would not be expected to cause a major change in the glass structure since there is no change in valence and only a small change in ionic radii. Yet, the substitution does have a dramatic effect on scintillation efficiency.

The accepted model for scintillation[3] involves three consecutive processes: ionization, energy transfer, and luminescence. Ionization, the generation of electrons and holes in the matrix, is the result of interactions between the incident radiation and the matrix. The subsequent motion of these electrons and holes through the matrix to an activator (Ce^{3+} in this case) constitutes the energy-transfer process. At the activator, the excitations recombine to produce luminescence.

The efficiency of this process can be affected in several ways. The creation of ionization is largely governed by the matrix, and, within the accepted models, small compositional variations can be expected to have little effect on the efficiency of this process. During the energy-transfer stage, the excitations can be trapped at charged defects in the matrix. This trapping will act to prevent or delay the excitation of, and eventual recombination at, the activator. Finally, the luminescence can be quenched by interaction between matrix energy states and those of the activator (non-radiative decay).

By examining light-generating mechanisms in a simple compositional series, it may be possible to discern if scintillation in cerium-activated glasses is dominated by any one of the three processes described above. Changes in charge transport or electron and hole mobility through the matrix to the activator will be indicated by changes in time or energy-dependent phenomena, such as scintillation lifetime. Direct excitation of fluorescence from the Ce^{3+} in the glass will indicate changes in the activator and its immediate environment and the relationship of activator energy states to those of the matrix.

EXPERIMENTAL

Glasses of the composition $[20Li_2O]\bullet[15AO]\bullet[64.4SiO_2]\bullet[0.6Ce_2O_3]$ (where A is Mg, Ca, Sr, or Ba) were prepared. In [I], results of studies on these glasses via scintillation efficiency at room temperature, scintillation lifetime, front-face fluorescence, thermoluminescence, Raman spectroscopy, and atomic-force microscopy were reported. Here, we report the results of scintillation-efficiency measurements at reduced temperatures, improved thermoluminescence studies, and scanning electron microscopy (SEM). The scintillation lifetime provides information on the mechanisms by which energy is transferred to the activator (Ce^{3+}). Raman spectroscopy provides information on the overall glass structure, giving the most information on the degree of disorder of the silica network[4]. Front-face fluorescence measurements provide indirect evidence as to the relative energies of the Ce^{3+} states and the relationship of these electronic states to other energy levels of the glass matrix (i. e., the ultraviolet [UV] band edge and Ce^{4+} states); direct experimental verification of the relationships between these energy states is difficult to obtain because the Ce absorption is so large that it masks all the other parts. Microstructural analyses are necessary to ensure that the data from other measurements are not misleading due to the presence of multiple phases.

High-purity raw materials[a] were dry mixed, melted in Pt crucibles at 1370 to 1380°C in a reducing atmosphere for 2 h, and cooled by sandwiching between water-cooled, platinum-coated nickel plates.

Scintillation efficiency. For the room-temperature measurements, annealed samples were mounted in an integrating hemisphere at the end of a photomultiplier tube (PMT)[5]. Each sample was surrounded by immersion oil to reduce the effects of internal reflection and sample-to-sample geometry effects. A ^{137}Cs gamma-ray source was used to excite the sample. A Compton-backscattering coincidence method was used to ensure that only those scintillation events that deposit a given amount of energy were counted. For the low-temperature measurements, the sample was placed on a cold finger with an integral heater; temperature was controlled by a Lakeshore 93C controller. A fused-silica rod served as a light pipe to carry the luminescence through the vacuum envelope and to the PMT. Scintillation-efficiency measurements at room temperature were used to calibrate the results at low temperature.

Thermoluminescence. The sample was placed on a cold finger with an integral heater and a platinum-resistance thermometer in a vacuum chamber. The sample was preannealed by heating to 250°C. It was then cooled to the exposure temperature. At the exposure temperature, the sample was illuminated with 324-nm ultraviolet light carried into the sample chamber via an optical fiber. For the experiments conducted here, the UV exposure was 5 min. The sample was heated at a linear rate to the final temperature. A photon-counting photomultiplier tube deliverd the signal to recording apparatus.

Microscopy. Samples were etched using a 2% nitric acid, 2% hydrofluoric acid solution for 25 sec. After etching, the sample was washed with deionized water, dried with pressurized fluorocarbon, and gold-coated for SEM analysis.

X-ray diffraction. X-ray powder diffraction measurements were performed from 5° to 75° 2θ in

---

[a] Magnesium oxalate - Johnson Mathey Puratronic; Barium carbonate - Electronic Space Products International; Calcium carbonate - EM [FO-Optipure]; Strontium carbonate - Johnson Mathey Puratronic; Silicon dioxide - EM [FO-Optipure]; Cerium acetate -Johnson Mathey REacton; Lithium carbonate - Noah Technologies

0.04° steps with an integration time of 2 sec using a Cu K_{α} x-ray tube.

RESULTS

Scintillation efficiency. The gamma-ray-induced scintillation efficiency at 296 K, 190 K, and 84 K are shown in Figure 1 as a percentage of that for NaI together with the ratio of the scintillation efficiencies at the two temperatures. It is seen that the amount of light produced varies by more than a factor of 2.5 at each temperature. This is a large effect. A change this large should be reflected in changes in other, energy-transfer-related phenomena in the glasses. At each temperature, the ratio of the efficiency to that at room temperature is approximately constant.

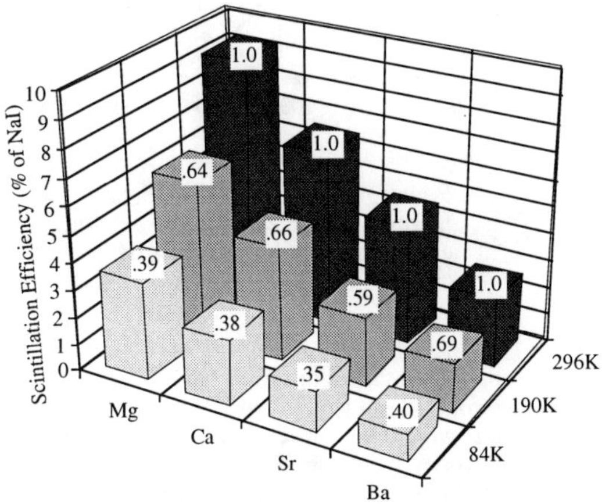

Figure 1. Scintillation efficiency of alkaline-earth glass series at 296 K, 190 K, and at 84 K as a percentage of that of NaI (left). The ratio of the efficiency to that at 296 K is shown at the top of the column.

Scintillation lifetime. The scintillation lifetime data are shown in Figure 2.

Thermoluminescence. The thermoluminescence measurements for exposure temperatures of -150°C and 20°C are shown in Figures 3 and 4. The spectra were normalized to unit amplitude for the larger peak.

Microscopy. The SEM measurements give information about the extent of phase separation in this series of glasses. Phase separation can confuse the interpretation of the scintillation-efficiency issue by mixing different phenomena in different phases. It could also be a primary cause of the differences in scintillation efficiency by affecting energy-transfer mechanisms. The SEM data (Figure 5) show distinct differences in the nature of the phase separation in the various compositions. These SEM measurements are consistent with the AFM results reported in [I].

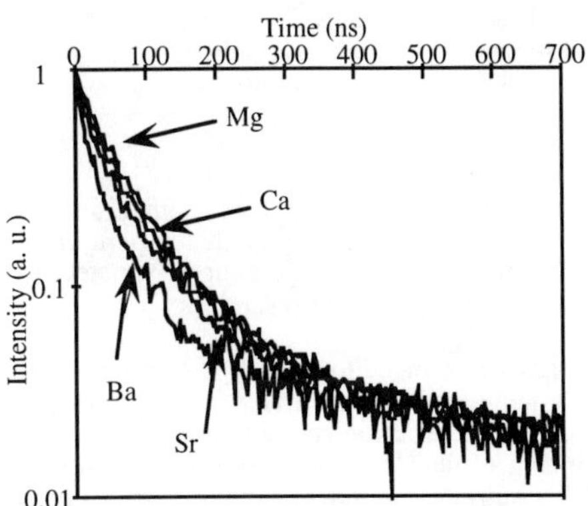

Figure 2. Luminescence lifetimes for alkaline-earth series.

The x-ray powder diffraction results for the Mg sample indicated that approximately 15% was crystalline, tentatively identified as cristobalite and CeO_2. The Ca, Sr, and Ba samples appeared to be amorphous.

DISCUSSION

It is clear that the scintillation efficiency significantly changes with the atomic number of the alkaline earth in an otherwise identical compositional series. By itself, this variation is surprising and reflects a corresponding variation in the energy-transfer mechanisms

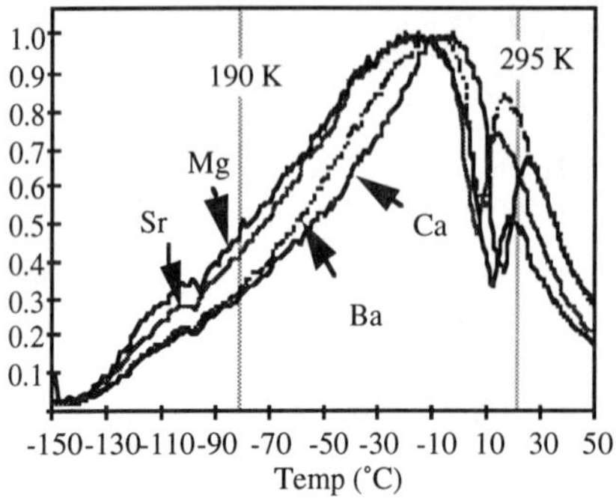

Figure 3. Relative low-temperature thermoluminescence of the alkaline-earth series. Temperatures at which scintillation efficiency was measured are indicated.

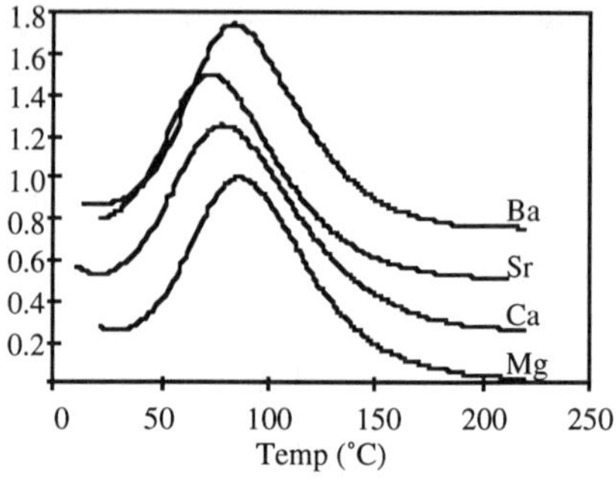

Figure 4. Relative high-temperature thermoluminescence of the alkaline-earth series. Curves are offset for clarity.

described in the introduction. No reason exists to suggest that the ionization step is being affected by varying the alkaline earth. The energy transfer from a fast charged particle is, to first order, proportional to the material density. The formula weight of these glasses only varies by approximately 35%. A 35% increase in ionization density is too small to account for the 3-fold decrease in scintillation efficiency. Thus, it is in the transport and luminescence steps that the difference is arising.

Electron transport in a glass is dominated by hopping conduction. Traps are charged structures that can capture electrons or holes. Shallow traps will slow the transport to the activator and result in longer components to the luminescence lifetime. Deep traps may delay the excitations long enough for them to recombine or for them to be no longer associated with the event.

Luminescence occurs when the electron and hole recombine on a Ce^{3+} ion to form an excited ion, which subsequently decays by the emission of light. Quenching of the luminescence occurs when the ion loses the energy without the emission of light. This can occur via direct thermal quenching between the activator ground and excited electronic states or overlap between the activator excited state and extended energy states of the matrix. Because quenching represents an additional pathway to excited state decay, it will shorten the luminescence lifetime.

Electron transport.

At each temperature, the ratio of the scintillation efficiency to that at room temperature is approximately constant for each sample. This variation in efficiency with temperature is consistent with traps dominating the scintillation efficiency if the traps are at approximately the same energy in all the samples and only the number of traps varies between samples.

The low-temperature thermoluminescence data provide information about the environment immediately adjacent to the Ce^{3+} activator; the high-temperature data provide information about the matrix further from the activator ion. The majority of the electron and hole transport is in the bulk of the material, away from the immediate influence of the Ce^{3+} ions. For this transport, the traps that give rise to the high-temperature peaks will control the mobility. The differences in the traps near the ions as characterized by the 10-30°C peak could explain the wide range of

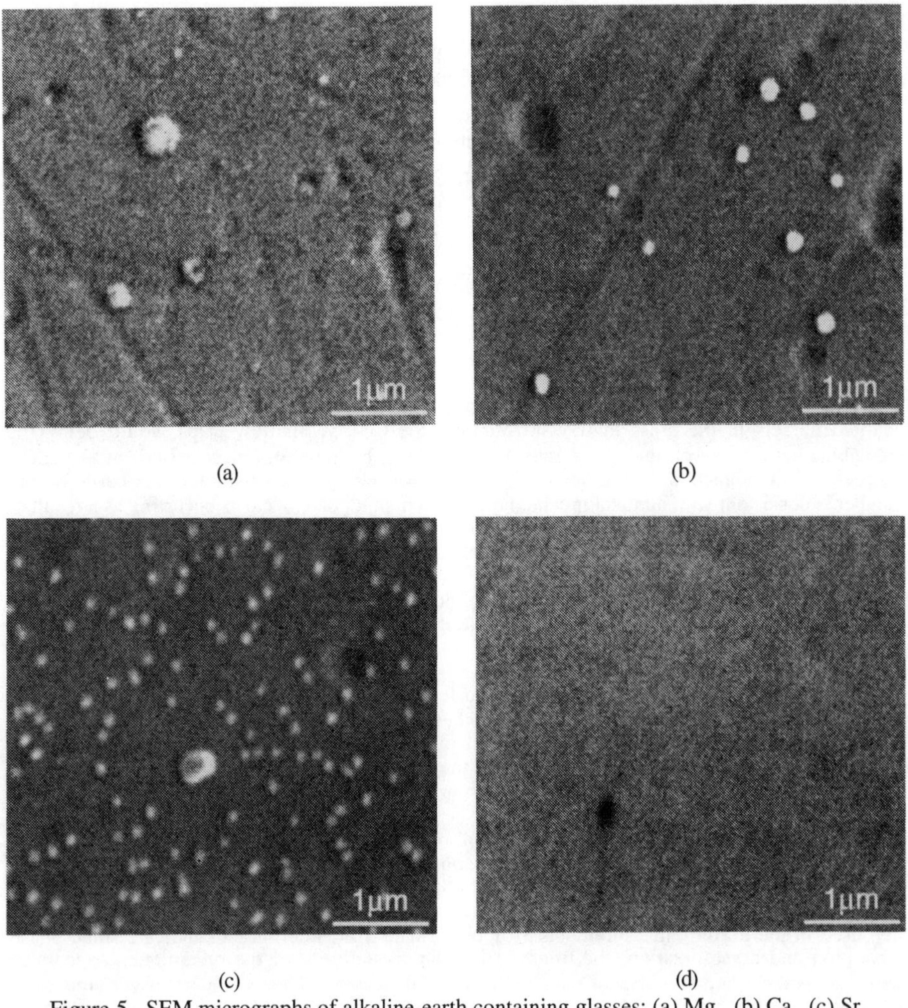

Figure 5. SEM micrographs of alkaline-earth containing glasses: (a) Mg, (b) Ca, (c) Sr, and (d) Ba.

scintillation efficiency for the alkaline-earth series at room temperature; it is less credible to believe that the relatively small differences in these traps significantly affect the efficiencies at 84 K and 190 K. On the other hand, the fact that the scintillation efficiency at room and low temperatures vary in the same way across the series suggests that electron-hole mobility differences are controlling the efficiency.

<u>Luminescence.</u>

Fluorometry measurements [I] indicated an interaction between the UV band edge and excited Ce^{3+} states for Ba. Smith and Cohen[6] measured the absorption of sodium-silicate glasses melted under oxidizing conditions with alkaline earths. They found an absorption band centered at approximately 5.8 eV (approximately 215 nm), which they noted to be increasing in intensity with the atomic weight of the alkaline earth. At high concentrations, this band would appear as a reduction of the UV bandgap. The room-temperature scintillation efficiencies of our glasses are consistent with quenching by interaction with this band. This is consistent with the lifetime data. On the other hand, if phonon-mediated quenching is the dominant mechanism, an increase in efficiency should be observed at lower temperatures; this increase should be greatest for the glasses which are most quenched. This is not observed.

In [I] the scintillation-light decay curves were fit with three lifetime components: a fast component of 5 to 10 ns, a moderate component of 45 to 60 ns, and a slow component of >400 ns. The abundance of the two slower components was found to vary with alkaline earth; the abundance of the fastest component is small but essentially constant. (The abundance is the total amount of light arising from a given lifetime component. For the long component, it is the amount of light that is released within 2 μsec.)

The slower components are expected to be controlled by thermal detrapping of excitations from charged defects in the glass matrix surrounding the Ce^{3+}. Shallow traps would delay the recombination for a short time; deep traps delay the recombination for longer periods of time. The decrease in abundance and lifetimes of the two slower components with alkaline-earth atomic number is consistent with a model in which increased nonradiative decay is occurring as a result of quenching. (See the increased early decay rate of the luminescence in Figure 2.) The nearly constant abundance of the fast component is unexplained.

Additionally, the Raman data discussed in [I] indicate no significant differences in the phonon spectra. This provides further evidence that lattice vibrations are not involved in quenching the luminescence.

The front-face fluorescence spectra are shown in [I]. The spectrum maximum move to longer wavelengths as the alkaline-earth atomic number is increased. This effect is consistent with quenching or absorption by the glass matrix as the UV-band edge decreases for higher-atomic-weight alkaline earths. On the other hand, this may be an unrelated consequence of the reduction of the UV bandgap.

The scintillation-light decay curves also yield information on the luminescence step. The decrease in abundance and lifetimes of the two slower components with alkaline-earth atomic number is consistent with a model in which increased nonradiative decay is a result of quenching. The nearly constant abundance of the fast component is unexplained but may be an artifact of the 3.4-ns time step used in the lifetime measurements.Structural. The SEM micrographs indicate some phase separation and crystallization; the trend is that the crystallinity of the precipitates appears to decrease as the atomic number of the alkaline earth increases. This is consistent with the x-ray diffraction results, which indicate a measurable crystallinity in the Mg-containing sample; the other glasses are x-ray amorphous.

SUMMARY

This work provides no clearcut answer for the observed variation in scintillation efficiency. The data are consistent with a hypothesis that luminescence is being quenched via interaction of Ce^{3+} excited states with an absorption band introduced by the alkaline earth. The data are not consistent with phonon-mediated or -assisted quenching. Another mechanism is required. The low-temperature thermoluminescent data show that the Ce^{3+} activator ions are surrounded by numerous shallow traps which are accessible at low temperature. If the absorption bands are directly accessible to these traps this can provide a mechanism for quenching with no or small temperature dependence.

The presence of the crystal-like inclusions in the lighter alkaline earths is interesting but provides no link with the scintillation efficiency. Scintillation efficiency measurements at lower temperatures than those used here and absolute quantum efficiencies of fluorescence are required to further develop models of the behavior of this glass system.

ACKNOWLEDGEMENTS

This work was supported, in part, by the Department of Energy, NN-20. Pacific Northwest Laboratory is operated by Battelle Memorial Institute for the U. S. Department of Energy under Contract DE-AC06-76RLO 1830.

REFERENCES

1. R. J. Ginther, "Glass Scintillators," IRE Trans. Nucl. Sci., **NS-5**, 92-95 (1958).

2. M. Bliss, R.A. Craig, P.L. Reeder, D. S. Sunberg, and M. J. Weber, "RELATIONSHIP BETWEEN MICROSTRUCTURE AND EFFICIENCY OF SCINTILLATING GLASSES," Mat. Res. Soc. Symp. Proc. **Vol. 348**, 195-202 (1994).

3. A. Lempicki and A.J. Wojtowicz, "Fundamental Limitations of Scintillators," J. Lumin., **60** & **61**, 942 - 947 (1994).

4. S.A. Brawer and W.B. White, "Raman Spectroscopic Investigation of the Structure of Silicate Glasses. I. The Binary Alkali Silicates," J. Chem. Phys., **63**, 2421-2431 (1975).

5. M. Bliss, R.A. Craig, and P.L. Reeder, "Method for Measuring the Light Output of Scintillating Glass Shards," Nucl. Inst. and Meth., **A 345**, 95-98 (1994).

6. H. L. Smith and A. J. Cohen, "Absorption spectra of cations in alkali-silicate glasses of high ultra-violet transmission," Physics and Chemistry of Glasses, **4**, 173ff (1963).

RELATIONSHIP BETWEEN MICROSTRUCTURE AND EFFICIENCY OF LITHIUM SILICATE SCINTILLATING GLASSES: THE EFFECT OF REPLACING LITHIUM BY SODIUM

M. BLISS, R. A. CRAIG, D. S. SUNBERG, AND M. J. WEBER
Pacific Northwest National Laboratory, Box 999, Richland, WA 99352

ABSTRACT

Lithium-silicate glasses containing Ce^{3+} are scintillators. When ^6Li is used, the glasses are neutron detectors. When sodium is substituted for lithium the scintillation efficiency, under gamma-ray stimulation, increases and then decreases as a function of sodium content. Glasses made with all the lithium replaced by sodium exhibit only weak scintillation under pulsed X-ray excitation. In addition, the luminescence, under ultraviolet stimulation, is qualitatively different across this series. The results of microstructural analyses and low-temperature luminescence measurements are reported.

INTRODUCTION

Cerium-activated, lithium-silicate glasses constitute an important and interesting system of scintillators.[1] The accepted model for scintillation[2] involves three consecutive processes: ionization, energy transfer, and luminescence. Ionization, the generation of electrons and holes in the matrix, is the result of interactions between the ionizing radiation and the matrix. The subsequent motion of these electrons and holes through the matrix to an activator (Ce^{3+} in this case) constitutes the energy-transfer process. At the activator, the excitations recombine to produce luminescence.

Several things can affect the efficiency of this process. The creation of ionization is largely governed by the matrix, and, within the accepted models, materials variations can be expected to have a small effect on the efficiency of this process. During the energy-transfer stage, the excitations can be trapped at charged defects in the matrix. This trapping will act to prevent or delay the excitation of and eventual recombination at the activator. Finally, the luminescence can be quenched by the interaction between matrix energy states and those of the activator.

By examining light-generating mechanisms in a simple compositional series, it may be possible to discern if scintillation in cerium-activated glasses is dominated by any one of the three processes described above. Changes in charge transport or electron and hole mobility through the matrix to the activator will be indicated by changes in time or energy-dependent phenomena, such as scintillation lifetime. Direct excitation of fluorescence from the Ce^{3+} in the glass indicates changes in the activator and its immediate environment as well as the relationship of activator energy states to those of the matrix. The high cerium content of these glasses makes direct measurement of peaks in the ultraviolet/visible (UV-vis)absorption spectrum extremely difficult.

EXPERIMENTAL

Glasses of the composition $(25\text{-}x)Li_2O{\bullet}xNa_2O{\bullet}74.4SiO_2{\bullet}0.6Ce_2O_3$ were prepared. These glasses were studied via scintillation efficiency, scintillation lifetime, front-face fluorescence, Raman spectroscopy, and scanning electron microscopy (SEM). High-purity raw materials[a] were dry mixed, melted for two hours in platinum crucibles at 1370 to 1380°C in a reducing atmosphere, and cooled by pouring onto a water-cooled, platinum-coated nickel plate. Use of a reducing atmosphere minimizes Ce^{4+}, which will absorb light emitted by the Ce^{3+}.

Scintillation efficiency. For the room-temperature measurements, annealed samples were mounted in an integrating hemisphere at the end of a photomultiplier tube (PMT). Each sample was surrounded by immersion oil to reduce the effects of internal reflection and sample-to-sample geometry effects. A ^{137}Cs gamma ray source was used to excite the sample. A Compton-backscattering coincidence method was used to ensure that only those scintillation events that deposit a given amount of energy are counted[3]. For the low-temperature measurements, the sample was placed on a cold finger. A fused-silica rod served as a light pipe to carry the luminescence through the vacuum envelope and to the PMT.

Scintillation lifetime. Representative samples were excited with 100-ps pulsed 30-keV X-rays generated by an optically excited X-ray tube. The scintillation was measured with a microchannel PMT feeding a standard time-to-amplitude convertor arrangement[4].

Raman Spectroscopy. Raman spectra were collected from annealed and polished samples using argon-ion laser light (514.5 nm) and a double monochronometer instrument. Spectra were gathered from 50 cm^{-1} through 1450 cm^{-1} in 1.0 cm^{-1} increments with a 2.5-sec integration time.

Front-face fluorescence. The Raman samples were mounted in a holder (blackened to minimize the effect of light passing through the sample and reflecting off the holder) at the center of a 6-in integrating sphere. These were excited with 325-nm radiation from a HeCd laser arranged such that the light just grazes the front face of the sample. Optics between the exit port of the integrating sphere and the spectrometer slit match the aperture of the spectrometer.

Microscopy. Polished samples were etched using a 2% nitric acid, 2% hydrofluoric acid solution for 25 sec. After etching, the sample was washed with deionized water, dried with pressurized fluorocarbon, and gold-coated.

RESULTS AND DISCUSSION

Scintillation efficiency. The gamma-ray-induced scintillation efficiency at 295 K and 84 K are shown in Figure 1a as a percentage of that for NaI. Figure 1b shows the ratio of the scintillation efficiencies at the two temperatures. The amount of light produced varies from more than 8% of NaI to less than 1% of NaI. This is a large effect. A change of this magnitude should be reflected in changes in other, energy-transfer-related phenomena in the glasses. For samples having low efficiency, the increase in efficiency for low temperature is consistent with a thermal-quenching phenomenon.

Scintillation lifetime. The results of the scintillation-lifetime measurements are shown in Figure 2 for the first 200 ns. The measurements were carried out to approximately 8 μs, but the principal difference is seen in the 0-200 ns regime. The luminescence decay over the first few tens of nanoseconds varies, for the glasses measured, in the same way as the light output. This is consistent with quenching of the luminescence by a fast non-radiative decay mode. The lack of significant differences for longer times indicates that traps are not a significant cause of the large variation in scintillation efficiency with composition.

---

[a] Magnesium oxalate - Johnson Mathey Puratronic; Barium carbonate - Electronic Space Products International; Calcium carbonate - EM [FO-Optipure]; Strontium carbonate - Johnson Mathey Puratronic; Silicon dioxide - EM [FO-Optipure]; Cerium acetate -Johnson Mathey REacton; Lithium carbonate - Noah Technologies

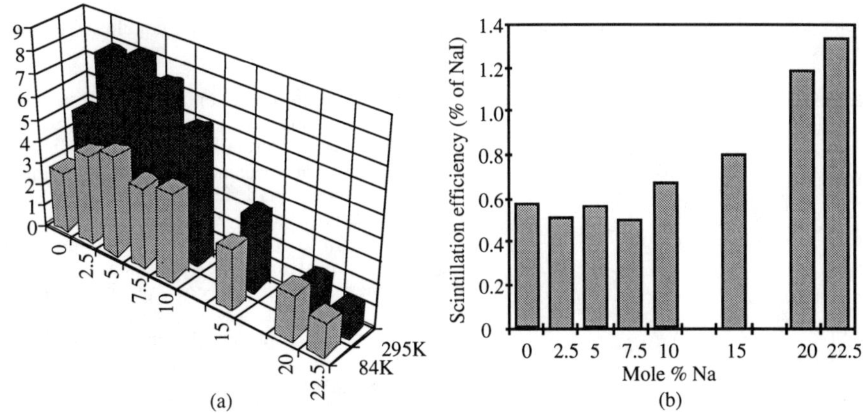

Figure 1. Scintillation efficiency of Li/Na glass series at 295K and at 84K as a percentage of that of NaI (a) and ratio of light output at 84K to that at 295K (b).

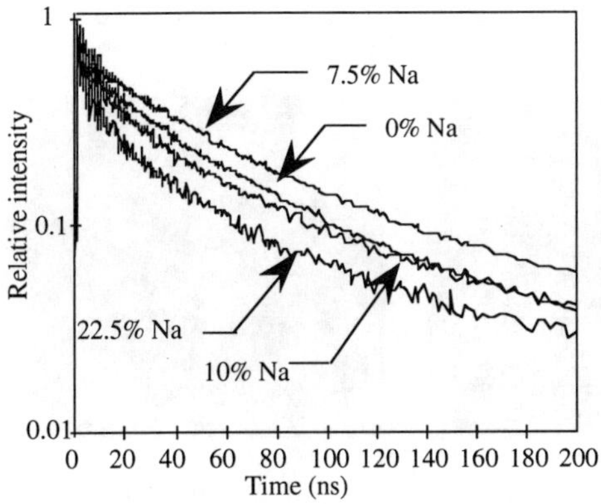

Figure 2. Scintillation lifetime measurements for selected glasses in the Li/Na series at 295 K.

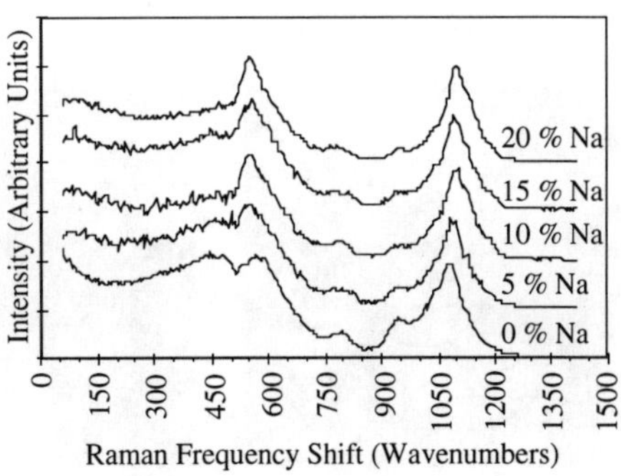

Figure 3. Raman spectra for the Li/Na-series. The intensities are normalized to the 1070/1090 cm^{-1} peak.

Raman Spectroscopy. The Raman data, normalized to the 1070- to 1090-cm^{-1} Si-O stretching band, are shown in Figure 3. The spectra are consistent with those obtained by others.[5,6] The peak shapes vary slightly with composition and do not have any discontinuities due to the mixing of the two alkali oxides. The behavior of the Si-O stretching bands in the 900- to 1300-cm^{-1} regime shows a gradual decrease in the ~960 cm^{-1} band as lithium is replaced by sodium, as reported by Brawer and White[5].

Front-face fluorescence. The front-face fluorescence spectra, normalized to unit amplitude at the fluorescence peak maximum, are shown in Figure 4. The spectra maxima move to longer wavelengths as the sodium content is increased. This is a small effect and does not clearly identify the cause of the large change in scintillation efficiency.

Microscopy. The SEM measurements indicate the extent of phase separation and crystallization in the various glasses. Phase separation can confuse the interpretation of the scintillation-efficiency data because different phenomena could dominate in different phases. The SEM micrographs (Figure 5) show distinct differences from sample to sample. The uniform distribution of the minor phases is consistent with their formation on cooling. No clear-cut relationship exists between the microstructure and the variation in scintillation efficiency.

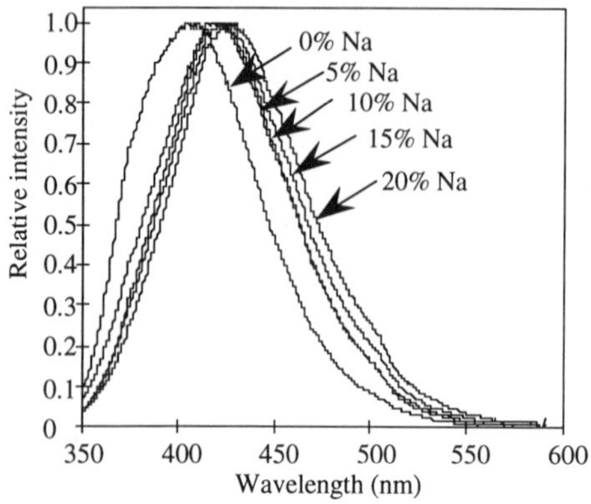

Figure 4. Front-face fluorescence spectra for varying sodium content. Spectra are normalized to unit peak amplitude.

SUMMARY

It is clear that the scintillation efficiency dramatically changes with the sodium content in this series, which is otherwise identical in composition. By itself, this variation is surprising and reflects a significant change in the energy-transfer mechanisms described in the introduction. There is no evidence in the data

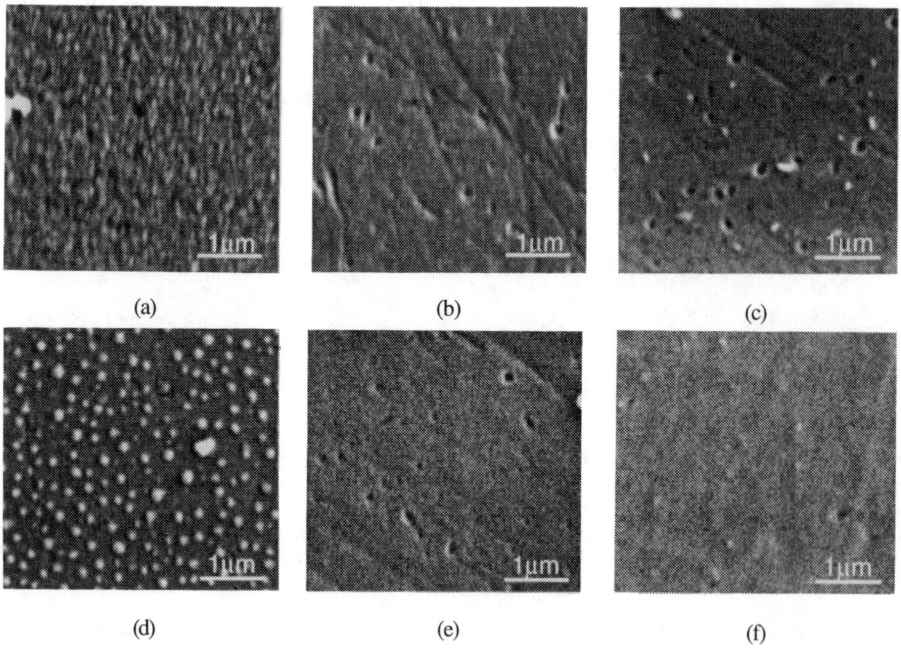

Figure 5. Etched SEM micrographs of selected sodium series glasses. (a) 25 mole % Li_2O; (b) 20 mole % Li_2O, 5 mole % Na_2O; (c) 15 mole % Li_2O, 10 mole % Na_2O; (d) 10 mole % Li_2O, 15 mole % Na_2O; (e) 5 mole % Li_2O, 20 mole % Na_2O; (f) 25 mole % Na_2O.

to suggest that the ionization step (electron-hole pair generation) is being affected by varying the sodium content. Replacing lithium with sodium increases the overall Z of the glass only slightly and does not dramatically change the phonon spectra.

The Raman spectra of the glasses do not indicate any anomalous behavior. White, in his study of phase-separated glasses, concluded that the Raman spectra are characteristic of structural units smaller than the scale of the phase separation averaged over the entire sample.[7] Although the SEM micrographs indicate some phase separation and crystallization, no strong correlation exists with the scintillation efficiency at room temperature or low temperature. The dissimilar behavior of the scintillation efficiency at room and low temperatures indicates that electron-hole mobility differences do not dominate the scintillation behavior. These data imply that the scintillation efficiency is affected by changes on a very fine scale, close to the cerium atoms.

The luminescence step, light finally being emitted by the cerium, most probably dominates the room-temperature scintillation behavior. Sigel[8] has shown that the UV absorption edge moves to longer wavelengths when alkali is added to silica glass and that the increase is greater for sodium than lithium. A decrease in the energy of the UV-absorption edge has two possible effects: 1) the light emitted by the activator (cerium) can be more readily self-absorbed, and 2) the excited states of the activator will be closer in energy to the extended states of the matrix (UV absorption edge). The latter of these will result in quenching of the luminescence via non-radiative decay. This phenomenon is consistent with the lifetime data and with the temperature dependence of the scintillation efficiency.

A reasonable hypothesis consistent with these data and the luminescence lifetime data is that the glasses fall into two categories: 1) those with up to 5 mole % sodium substituted for lithium and 2) the glasses containing sodium in greater than 5 mole %. In the first group of glasses, the scintillation efficiency increases because of structural changes in the immediate vicinity of the activator. This is consistent with compositional effects observed in rare earth doped laser glasses.[9] For the second category of glasses, those with greater than 5 mole % sodium, the extended matrix states are interacting with the cerium excited states and allowing quenching to occur. The room temperature scintillation efficiency is dominated by changes in the UV absorption edge.

ACKNOWLEDGEMENTS

This work was supported, in part, by the Department of Energy, NN-20. Pacific Northwest Laboratory is operated by Battelle Memorial Institute for the U. S. Department of Energy under Contract DE-AC06-76RLO 1830.

REFERENCES

1. R. J. Ginther, "Glass Scintillators", IRE Trans. Nucl. Sci., **NS-5**, 92-95 (1958).

2. A. Lempicki and A.J. Wojtowicz, "Fundamental Limitations of Scintillators", J. Lumin., **60** & **61**, 942 - 947 (1994).

3. M. Bliss, R.A. Craig, and P.L. Reeder, "Method for Measuring the Light output of Scintillating Glass Shards", Nucl. Inst. and Meth., **A 345**, 95-98 (1994).

4. S.C. Blankespoor, S.E. Derenzo, W.W. Moses, M. Ito, and K. Oba, "Characterization of a Pulsed X-ray Source for Fluorescent Lifetime Measurements", IEEE Trans. Nucl. Sci., **42**, (1994).

5. S.A. Brawer and W.B. White, "Raman Spectroscopic Investigation of the Structure of Silicate Glasses. I. The Binary Alkali Silicates", J. Chem. Phys., **63**, 2421-2431 (1975).

6. D.W. Matson, S.K. Sharma, and J.A. Philpotts, “The Structure of High-Silica Alkali-Silicate Glasses. A Raman Spectroscopic Investigation”, J. Non-Cryst. Sol., **58**, 323-352 (1983).

7. W.B. White, “Investigation of Phase Separation by Raman Spectroscopy”, J. Non-Cryst. Sol., **49**, 321-329 (1982).

8. G.H. Sigel, Jr., “ Vacuum Ultraviolet Absorption in Alkali Doped Fused Silica and Silicate Glasses”, J. Phys. Chem. Solids, **32**, 2373-2383 (1971).

9. Y. Ishii, K. Arai, H. Namidawa, M. Tanaka, A. Negishi, and T. Handa, “ Preparation of Cerium-Activated Silica Glasses: Phosphorus and Aluminum Codoping Effects on Absorption and Fluorescence Properties”, J. Amer. Ceram. Soc., **70**, 72-77 (1987).

Optical and Magneto-Optical Applications

FARADAY EFFECT IN LANTHANIDE-DOPED OXIDE GLASSES

Jeffrey T. Kohli
Corning Incorporated, SP-FR-5-1, Corning, NY 14831

INTRODUCTION

The lanthanides, often referred to as rare earths, are important components in many materials, including: magnets, pigments, polishing compounds, phosphors, lasers, and inorganic glasses. Despite the fact that they are thought of as being rare, the elements cerium and neodymium are as abundant in nature as lead and tin, while even the less common lanthanides are more abundant than the platinum group metals, gold or silver. An interesting footnote regarding the lanthanides deals with the location of the series in the periodic table. Of course, the term lanthanides is derived from the element lanthanum, which directly precedes the lanthanide series. However, both the lanthanide and actinide series are placed directly below the main body of the Mendeleev periodic table, so that the chart has a simple, rectangular form.

The lanthanides occupy a unique place in the history of science as well as in the periodic table. The formal discovery of the lanthanides occurred in 1787 in the town of Ytterby, Sweden, where C.A. Arrhenius uncovered a dark mineral ore. In 1794, Johann Gadolin separated an oxide from such a mineral. The mineral was later named gadolinite, and the oxide named yttria. Similarly, most of the lanthanides were named for the discoverer of the mineral or the location in which they were found. For example, the names of the elements yttrium, ytterbium, erbium, and terbium are all derivatives of the name Ytterby. Holmium and europium were named after Stockholm and Europe. The name thulium is derived from the word Thule, the former name of Scandinavia. Gadolinium, of course, was named after Gadolin. Other lanthanides were given Greek names, such as praseodymium, meaning green twin, and neodymium, translated as new twin. The word lanthanum means to lie hidden; perhaps impetus for the term rare earth. [1]

The history of inorganic glasses containing the lanthanides extends at least as far back in time as the 1890's. G.P. Drossbach [2] applied for a patent using rare earth oxides as decolorants in 1896. In the early 1900's the Moser Glass Works in Czechoslovakia was renowned for its production of beautiful glasses containing lanthanides, particularly neodymium. In 1932, Johannes Löffler [3] published some of the first technical discussions regarding the use of lanthanides as both colorizers and decolorizers. Also of early note, is Kreidl's [4] review of rare earths in the ceramics and glass industries up to the 1940's. More recently, Weyl's [5] book pertaining to colored glasses gives a historical account of the use of lanthanides in glass. Today, the most common uses for lanthanides in the glass industry include: lasers, Faraday rotators, optical fiber amplifiers, certain optical glasses, art glass, and of course, polishing compounds based on cerium oxide.

Several interesting optical effects were observed in the 1800's, including: the electro-optic, acousto-optic, and magneto-optic effects. In the last 30 years these effects have received a great deal of attention. The field of magneto-optics includes the following optical effects: Cotton-Mouton, Faraday, Kerr, Voigt, Zeeman, and Inverse Zeeman [6]. In 1845, Michael Faraday observed that a piece of glass placed in a magnetic field becomes optically active, such that the plane of polarization of light transmitted through the article is rotated.

The present paper pertains to the Faraday effect observed in oxide glasses containing high concentrations of lanthanide ions. It will be shown that magnetic measurements alone cannot be used to predict the magneto-optical properties of these and other optical materials containing high concentrations of lanthanide ions. Moreover, the relative magnitudes of the magnetic and magneto-optical coefficients are virtually independent of the identity of the diamagnetic host glass, irrespective of glass structure (if the valence of the rare earth ion is constant). Data obtained by the author is compared with data found in the literature.

NATURE OF THE FARADAY EFFECT

Although Michael Faraday discovered optical rotation under the influence of a magnetic field using a piece of glass, the effect he observed is not limited to ordinary glasses. Moreover, the Faraday effect is not confined to the visible region of the electro-magnetic spectrum. The Faraday effect has been observed in a variety of materials in the ultraviolet, infrared, and even the microwave [7] and X-ray [8] regions of the spectrum.

From Maxwell's theory, we know that light is composed of both right and left circularly polarized waves. These waves are superposed, are of the same frequency, and travel at the same speed in a material in the absence of an applied magnetic field. Absorption of light occurs in materials at specific resonance frequencies depending upon the composition of the material and the electronic structure of the constituent atoms. If a magnetic field is imposed on the material, then two closely-spaced resonance frequencies are created by a Zeeman-effect splitting of the original resonance. These frequencies correspond to right and left circularly polarized light [6]. The type of transition which occurs within the ions of the material determines if the clockwise or counterclockwise circularly polarized wave travels faster than the other, thereby leading to a net rotation of the plane of polarization in one sense or the other. After the right and left components of light exit a material placed within a magnetic field, they recombine to form a linearly polarized wave whose plane of polarization has been altered from that of the original beam. Paramagnetic transitions cause rotations in one sense and diamagnetic transitions cause opposite rotations. When the angle of rotation is clockwise (positive), as light propagates away from the viewer in the direction of the magnetic field, the rotation is caused by a diamagnetic transition. If the rotation is in the opposite direction (negative), then it is a paramagnetic rotation [9]. If a material contains both paramagnetic and diamagnetic ions, the outgoing beam is usually dominated by paramagnetic contributions, unless the paramagnetic ions in the glass have low probability transitions, in which case a diamagnetic rotation may be the net response. Some materials which have been defined as paramagnetic with the use of a magnetic susceptibility balance may have an optical rotation which is in the opposite direction to a paramagnetic rotation, according to the definitions discussed above. This apparent paradox arises from the fact that the probabilities of various transitions are unique; some leading to optical activity, others not. This phenomenon is discussed in greater detail later. The reader should note that the sign convention for the Faraday effect and magnetic susceptibility is reversed, i.e. the magnetic susceptibility of a paramagnetic material is positive, while the so-called paramagnetic Faraday rotation carries a negative sign.

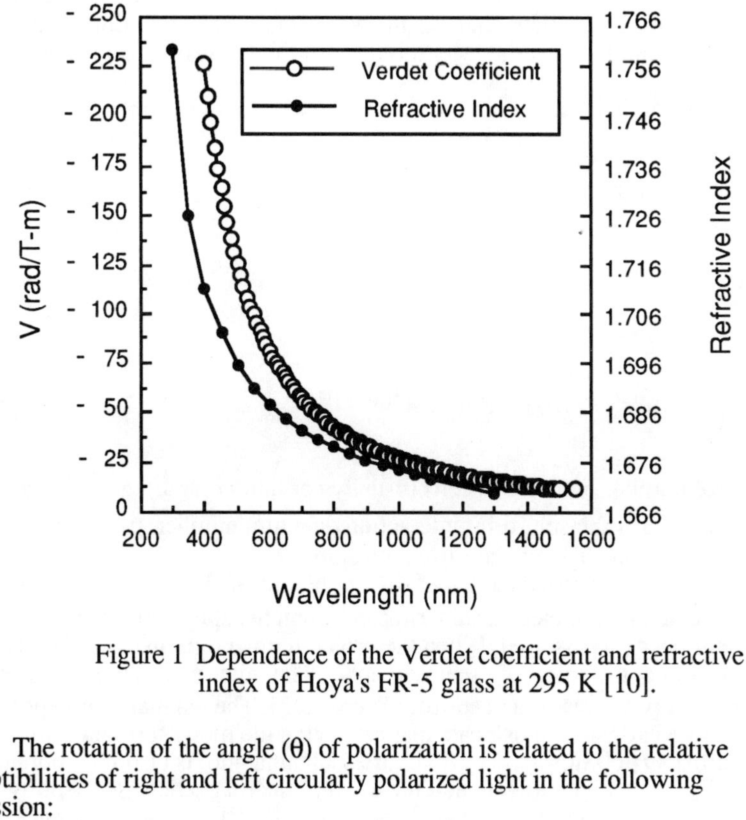

Figure 1 Dependence of the Verdet coefficient and refractive index of Hoya's FR-5 glass at 295 K [10].

The rotation of the angle (θ) of polarization is related to the relative susceptibilities of right and left circularly polarized light in the following expression:

$$\theta = \frac{4\pi}{\lambda}\left[\frac{(\bar{n}^2+2)^2}{\bar{n}}\right]\left(\chi_+ - \chi_-\right) L \qquad (1)$$

where λ is the wavelength of light, n is the refractive index , χ is the electric susceptibility of the polarized light, and L is the path length in the material [11]. Since this relationship is a function of the refractive index of a material, it clearly explains why the angle of rotation increases as the frequency of the light approaches a fundamental absorption edge (see Figure 1). The angle of rotation observed in a material under a given set of experimental conditions is expressed using the Verdet constant (V)

$$V = \theta/LH \cdot n \tag{2}$$

again θ is the angle of rotation, L is the path length of light in the material, H is the strength of the magnetic field, and n is the unit vector in the direction of light propagation within the material. Measurement of the Faraday effect is described elsewhere [12].

Since paramagnetism and the magneto-optical activity imparted to materials containing the lanthanides are related, it is worth reviewing some fundamental formulae. The classical form of the bound electron paramagnetic susceptibility is given by the relation:

$$\chi_{para} = \frac{N\, p_{eff}^2\, \mu_B^2}{3kT} \tag{3}$$

where

$$p_{eff} = g[J(J+1)]^{1/2} \tag{4}$$

and N is the number of atoms or ions (usually per unit volume), g is the Lande splitting factor, J is the spin-orbit interaction quantum number, μ_B is the Bohr magneton, and k and T have their usual meanings.

Figure 2 illustrates the relative susceptibilities of the trivalent lanthanide ions, represented by the theoretical term p_{eff}^2 when the spin-orbit interaction quantum number (J) is is used. When the orbital momentum quantum number (L) is equal to zero, i.e. the spin-only scenario, g=2. The other values used for g in the calculations of p_{eff}^2 were obtained from Weber [13]. The magnetic susceptibility of a series of rare earth aluminosilicate glasses having the molar formula $20R_2O_x$-$20Al_2O_3$-$60SiO_2$ (where R signifies a lanthanide) is also incorporated in Figure 2. The reader should note that the trend of the magnetic susceptibility of this series of glasses follows closely that of the term p_{eff}^2. These data have the same relative magnitudes as the susceptibilities of a series of rare earth sulfates presented by Van Vleck and others [14].

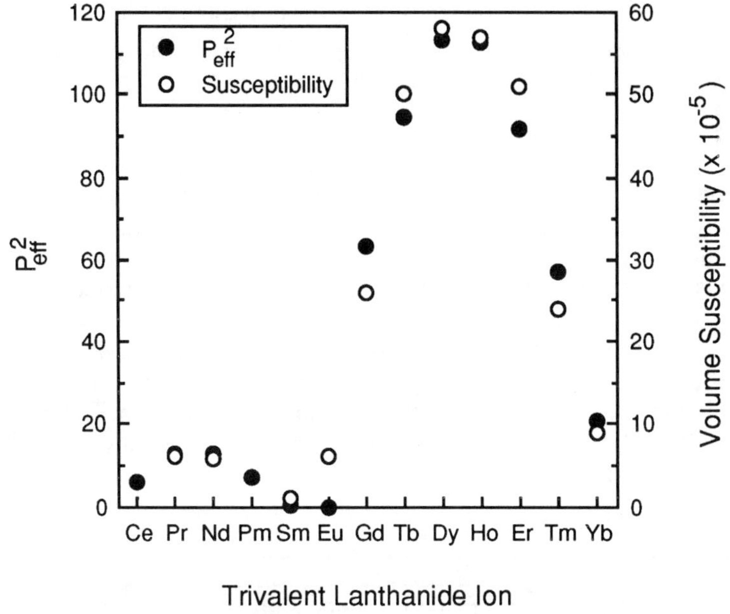

Figure 2 The dependence of the term p_{eff}^2 and the room temperature magnetic susceptibility of a series of glasses of the formula $20R_2O_x$-$20Al_2O_3$-$60SiO_2$ as a function of the lanthanide.

The paramagnetic behavior of the glasses represented in Figure 2 results from the unpaired electrons in the shielded 4f electronic suborbitals of the various lanthanide ions. The electrons in the 4f suborbitals of these ions are not significantly affected by their ligand fields. Van Vleck concluded that rare earth ions may be considered "free" ions in solids, because the unpaired 4f electrons are shielded so effectively by outerlying electrons. Thus, one may conclude that the relative magnetic susceptibilities of any similar series of materials containing the different rare earth ions (and no other paramagnetic ions) should have the form of the data in Figure 2.

The relationship between the paramagnetic Verdet (V_{para}) constant and the paramagnetic susceptibility was established by Van Vleck and Hebb in the early 1930's [15]. The following classical expression correlating the two quantities is:

$$V_{para} = \frac{4\pi^2\nu^2\chi_{para}}{g\mu_B ch} \sum_{a,b} \left(\frac{C_{ab}}{\nu^2 - \nu_{ab}^2}\right) \quad (5)$$

where υ is the frequency of light, c is the speed of light, h is Planck's constant, C_{ab} is a transition moment, and υ_{ab} is the frequency difference between the ground and excited states [11]. A derivation of the nature of paramagnetic optical rotations using time-dependent perturbation theory is given by Ballhausen [16].

The relationship in equation 5 demonstrates that the paramagnetic Verdet constant increases with the square of light frequency, i.e. V_{para} increases rapidly as wavelength decreases (see Figure 1). However, the terms in the summation indicate that near an absorption caused by an f→f or f→d transition in the lanthanides, an increase in the paramagnetic Verdet constant is expected. The magnitude of the increase is dependent upon the energy difference between the ground and the excited states, related by the term υ_{ab}.

Figure 3 illustrates the wavelength dependency of the Verdet constant in Hoya's FR-5 material. Unfortunately, this figure does not show the expected increase in the Verdet constant near the absorption at ca. 485 nm. The reason that a change in V is not exhibited in this diagram is because its measurement occurs too far from the absorption peak.

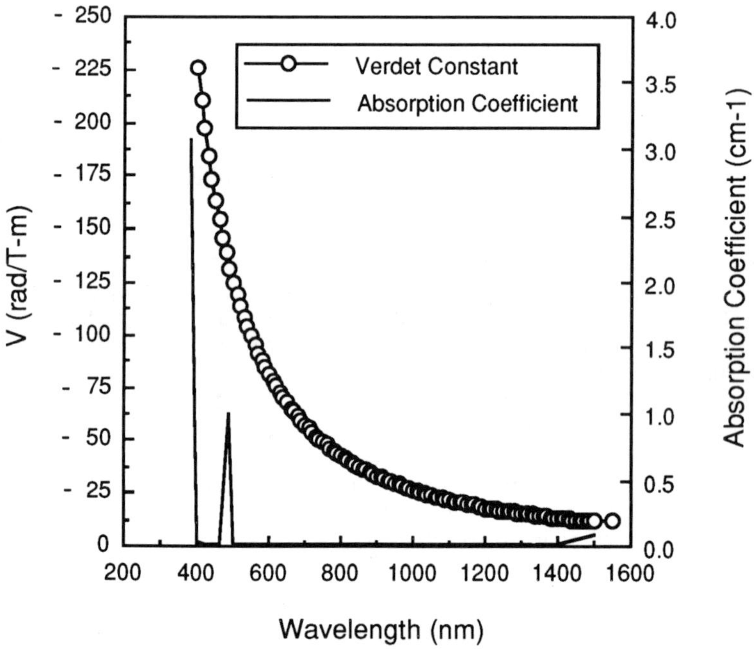

Figure 3 The wavelength dependence of the Verdet constant and absorption coefficient of FR-5 glass at 295 K [10].

The Verdet constant was measured at 480 and 490 nm in this particular data set; not between these two wavelengths. Collocott and Taylor [17,18] have demonstrated that a dramatic change in the Verdet constant is observed near an absorption caused by an electric-dipole transition. The rapid change in V, as one scans the spectrum, generally occurs over a very narrow wavelength region. In fact, near such an absorption peak, the Verdet constant takes the approximate form of the first derivative of the absorption curve, changing sign as the absorption region is traversed. This phenomenon is described by Jenkins and White [6]. For example, Collocott and Taylor have demonstrated that such a change occurs over a wavelength region of 1 nm (at 442.5 nm) in a Ho-doped glass [18]. The reader is reminded that the magnitude of such an effect is also temperature dependent.

FARADAY EFFECT IN VARIOUS LANTHANIDE-DOPED GLASSES

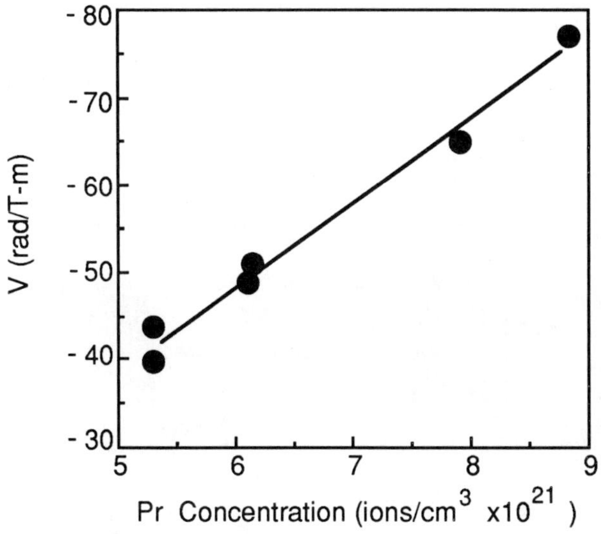

N (ions/cm^3 x10^{21})	Host Glass	Reference
5.30	Phosphate	Berger et al. [19]
5.30	Phosphate	NBS [13]
6.11	Galliogermanate	Lindquist [20]
6.16	Aluminogermanate	Lindquist [20]
7.93	Aluminosilicate	Kohli [21]
8.85	Aluminoborate	Spess [22]

Figure 4 Verdet constant versus praseodymium ion concentration for various glasses. V measured at 633 nm, 294-296 K.

The Verdet constants of glasses containing praseodymium are plotted in Figure 4 as a function of the concentration of the lanthanide ion in various host glasses. The Verdet constant increases linearly with the concentration of the rare earth ion in the glasses, as one would expect from equations 3 - 5. The reader should note that the host glass itself has little effect on the Verdet constant, because the ions in the glass, other than the lanthanides, are weakly diamagnetic. Incidently, at this level of "doping", praseodymium ions play a significant role in determining the structure and properties of these glasses.

The quantity V/N (Verdet constant divided by the concentration of the lanthanide ion in the glass) is useful for comparing lanthanide-doped glasses, because it clearly demonstrates that the Verdet constant in these glasses is dominated by the lanthanide ion concentration. V/N is plotted in Figure 5 as a function of the lanthanide in four different types of glasses. It can be seen that the trend in the data is somewhat different than the trend of the magnetic susceptibility data for similarly doped glasses (see Figure 2). It is of particular interest that some glasses measured as paramagnetic with the use of a susceptibility balance exhibit a diamagnetic magneto-optical rotation. Figure 6 further demonstrates this phenomenon, where the Verdet constant and volume magnetic susceptibility of a series of ytterbium aluminosilicate glasses is plotted as a function of ytterbium ion concentration.

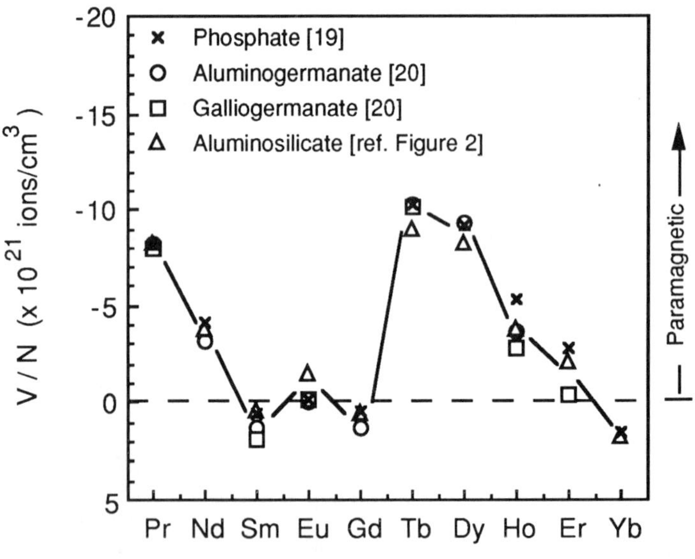

Figure 5 V/N versus lanthanide ion in various host glasses. V was measured with a 633 nm He-Ne laser at room temperature.

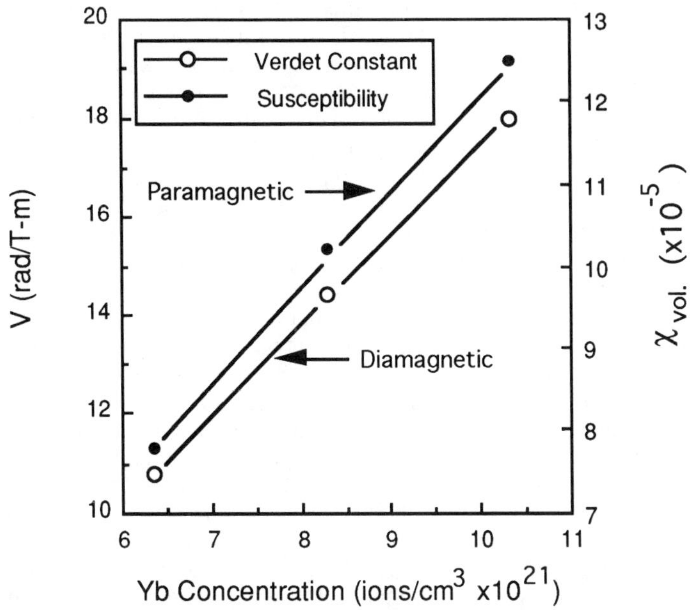

Figure 6 The dependence of Verdet constant and magnetic susceptibility on ytterbium ion concentration for a series of Yb aluminosilicate glasses (633 nm, 294-296 K).

Figure 6 may be confusing due to the problem of the sign convention referred to earlier in this manuscript. Careful examination reveals that as the Yb ion concentration increases, these glasses become more paramagnetic, due to the fact that each trivalent Yb ion contains one unpaired 4f electron. However, as the ytterbium ion concentration increases, the magneto-optical rotation becomes increasingly more diamagnetic.

DISCUSSION

It is noteworthy that although the glasses shown in Figures 2, 5 and 6 are all paramagnetic, the glasses containing samarium, gadolinium, and ytterbium exhibit diamagnetic rotations. The fact that the magnetic susceptibility and Verdet constant of a particular glass composition are not always of the same type (i.e. either diamagnetic or paramagnetic) and are not of the same relative magnitude, follows from the terms in the summation of equation 4. According to Weber [11], the magnitude of the Faraday rotation in materials containing lanthanides is dominated by $4f^n$ - $4f^{n-1}5d$ electric-dipole transitions in these ions. The energy separation between the ground and the excited states is related to the probability of a paramagnetic electric-dipole transition. Ions such as Gd^{+3} and Yb^{+3} have high-lying 5d bands relative to the energies of their ground 4f levels, yielding small rotations. The trivalent terbium ion, on the other hand, with a low-lying 5d band causes a considerable Faraday rotatory response. Even though the glasses containing ytterbium and gadolinium are magnetically paramagnetic, the probability of a $4f^n$ - $4f^{n-1}5d$ electric-dipole transition occurring in these rare earth ions is very low. Therefore it is not surprising that the diamagnetic interactions dominate the rotational response of the compositions containing gadolinium, samarium, and ytterbium. Weber has also stated that the Verdet constant of a material is related to the lifetime of the electronic transition in the rare earth ion [23].

The data in Figure 4 are consistent with the theory for the paramagnetic rotatory response of materials containing lanthanide ions. Since the concentration of the lanthanides in the glasses considered in the present manuscript are quite high, the diamagnetic contribution to the Verdet constant from the atoms which do not have unpaired electrons is typically negligible. In order for a material to exhibit a paramagnetic rotatory response, it must contain atoms with unpaired electrons. The measured rotation is the sum of the paramagnetic (-) and diamagnetic (+) contributions from the material. When lanthanide ions are present in a material, the Verdet constant simply increases with the concentration of the ions per unit of optical path distance. The Verdet constant of glass is constant, regardless of specimen orientation, because the material is isotropic. However, in many crystalline materials, the Verdet constant is dependent upon the orientation of the crystal relative to the magnetic field and the direction of light propagation.

CONCLUSIONS

The data shown in the present paper indicate that the magneto-optical properties of materials containing different trivalent lanthanide ions can be predicted from other materials containing these same ions. However, the magneto-optical response is not necessarily proportional to the magnetic susceptibility. Moreover, a material which has a paramagnetic susceptibility may have a diamagnetic rotatory response resulting from the spin-paired electrons in the atoms of the material, if a paramagnetic transition in the rare earth ion is not favored.

REFERENCES

1. T. Moeller, The Chemistry of the Lanthanides, Reinhold Publishing Corp., New York, 1963.

2. G.P. Drossbach, German Patent No. 103,441, 1896.

3. J. Löffler, "Chemische Entfärbung," Glastech. Ber., **10** [4] 204-211, 1931.

4. N.J. Kreidl, "Rare Earths," J. Am. Ceram. Soc., **25** [5] 141-143, 1942.

5. W.A. Weyl, Coloured Glasses, Society of Glass Technology, Second

Edition, Sheffield, England, 1976.

6. F.A. Jenkins and H.E. White, Fundamentals of Optics, Fourth Edition, McGraw-Hill Book Company, New York, 1976.

7. J.A. Weiss, Papers presented at the Conference on Magnetism and Magnetic Materials, Boston, MA, Oct. 16-18, 1956, Published by the American Institute of Electrical Engineers, T-91, 580-585, 1957.

8. D.P. Siddons, et al., "X-Ray Optical;l Activity and the Faraday Effect in Cobalt and Its Compounds," Phys. Rev. Lett., **64** [16] 1967-1970, 1990.

9. J. A. Davis and R. M. Bunch, "Temperature Dependence of the Faraday Rotation of Hoya FR-5 Glass," Appl. Opt., **23** [4] 633-636 (1984).

10. Hoya Corporation USA - Optics Division, Product Literature.

11. M. J. Weber, "Faraday Rotator Materials for Laser Systems," Laser and Nonlinear Optical Materials, Proc. SPIE Int. Soc. Opt. Engrs., **681** 75-90 (1986).

12. J.T. Kohli, "The Faraday Effect in Glasses Containing Rare Earths," Rare Elements in Glasses, Edited by J.E. Shelby, Key Engineering Materials, Trans Tech Publications, Switzerland, Vols. 94-95, 1994.

13. M. J. Weber, "Faraday Rotator Materials," Report for the U.S. Government (DOE) and University of California, Lawrence Livermore Laboratory, Livermore, 1982.

14. J. H. Van Vleck, The Theory of Electric and Magnetic Susceptibilities, Oxford University Press, London, 1932.

15. J. H. Van Vleck and M. H. Hebb, "On the Paramagnetic Rotation of Tysonite," Phys. Rev., **46** [10] 17-32 (1934).

16. C.J. Ballhausen, Introduction to Ligand Field Theory, McGraw-Hill Book Company, Inc., New York, 1962.

17. S.J. Collocott and K.N.R.Taylor, Magneto-Optical Properties of Erbium-Doped Soda Glass, J. Phys. C: Solid State Phys.**11** 2885-2893, 1978.

18. S.J. Collocott and K.N.R.Taylor, Magneto-Optical Properties of Praseodymium and Holmium-Doped Soda Glass," J. Phys. C: Solid State Phys. **12** 1767-1775, 1979.

19. S. B. Berger, C. B. Rubinstein, C. R. Kurkjian, A. W.Treptow, "Faraday Rotation of Rare Earth (III) Phosphate Glasses," Phys. Rev., **133** A723-A727 (1964).

20. B. H. Lindquist, NYS College of Ceramics at Alfred University, personal communication, 1991.

21. J. T. Kohli and J. E. Shelby, "Magneto-Optical Properties of Rare Earth Aluminosilicate Glasses, Phys. Chem. Glasses," **32** [3] 109-114, 1991.

22. M. S. Spess, NYS College of Ceramics at Alfred University, personal communication, 1991.

23. M. J. Weber, personal communication, 1991.

LANTHANIDE AND ACTINIDE DOPED GLASSES AS REFERENCE STANDARDS FOR DYE DOPED SYSTEMS

Edward J. A. Pope*, Matech, Westlake Village, CA, 91362
Art Hentschel, Spec Check, Torrance, CA 92635

ABSTRACT

Organic dye molecules are well known to be subject to chemical and optical bleaching damage, temperature instability, and other forms of optical degradation. Currently recognized methods of referencing rely upon fluorescent salt solutions, such as quinine sulfate. In this paper, optically-active lanthanide and actinide doped gel-glasses are compared as reference standards for dye doped polymers. Samples are subjected to continuous illumination by 254 nm UV radiation. While dye-doped polymers exhibited approximately 65 percent decline in fluorescence intensity after 96 hours of irradiation, glass samples and glass powder in resin showed no decline in fluorescence intensities.

INTRODUCTION

For thousands of years, colored glass has been utilized for both decorative and utilitarian purposes. In the cathedrals of Notre Dame and Chartres, the richly colored stained glass windows have delighted visitors and the faithful for over seven centuries. In addition to absorption, glasses can also exhibit fluorescence [1-3]. Lanthanide elements in glass, such as neodymium and erbium, have gained importance as fiber-optic amplifiers for telecommunications. Other lanthanides, such as cerium, europium, and samarium, exhibit strong fluorescent behavior in the visible part of the spectrum. Actinide elements, such as uranium and americium, also exhibit strong visible fluorescence. In this paper, the use of lanthanide and

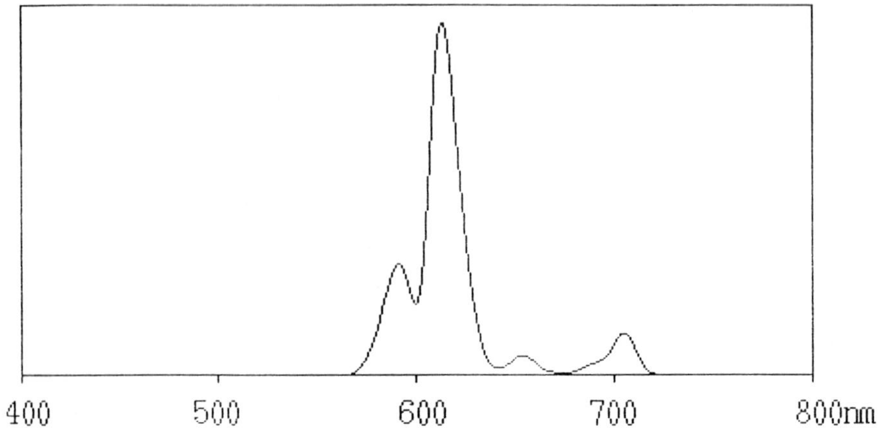

Figure 1: Fluorescence spectra of trivalent europium doped silica gel-glass.

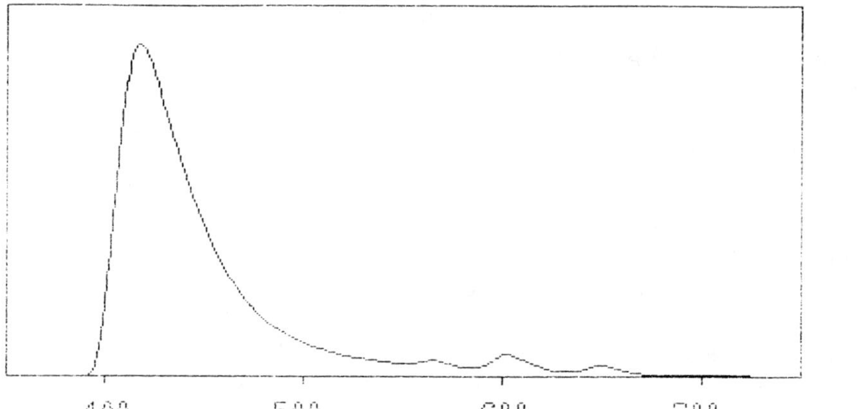

Figure 2: Fluorescence spectra of 0.2% cerium oxide doped borosilicate glass (Pyrex).

actinide doped gel-glasses as reference standards for dyes and fluorescence based equipment is demonstrated.

RARE EARTH AND ACTINIDE DOPED GELS AND GLASSES

A wide variety of optical and electronic materials can be prepared by the sol-gel route [4,5]. Sol-gel processing has been used to synthesize high purity silica for fiber preforms, Nd-doped silica for fiber-optic amplifiers, dye-doped silicate materials for solid-state dye lasers[6], non-linear optic devices[7], flat-panel displays[8], fiber-optic sensors[9],and photochromic materials[10]. Optically-active inorganic species can be successful doped into both melt-glass and gel-glass. In figure 1, the fluorescence spectra of trivalent europium doped silica gel is presented. The gel, while still porous, was heat treated to 800 C in air. The sample produces a strong, red fluorescence. In figure 2, the bright blue fluorescence of 0.2 % cerium oxide doped Pyrex glass, derived from the melt, is presented.

DYE DOPED POLYMERS AND GELS

Fluorescent dye molecules have been successfully utilized for dye laser systems for over two decades[11]. Progress in developing solid-state dye lasers, however, has been hampered by the rapid degradation of the dyes by optical bleaching. Utilizing dye-doped solid-state materials as fluorescence standards for callibrating measurement equipment has been hampered by the same problem. In figure 3, the fluorescence intensity is plotted versus irradiation time for four dye-doped PMMA samples and an optically active samarium and uranium doped glass. As can readily be seen the fluorescence intensity for all four dye doped samples decreased by approximately 60 percent. Similar behavior has been observed for prolonged UV irradiation of photochromic dye doped gels and polymers [10]. Chemical bleaching, by oxidation, has been observed for nonlinear optic dyes, such as methyl nitroanaline (MNA) [12].

STABILITY OF GEL-GLASS POWDERS IN POLYMERIC HOST

While the high long-term stability of lanthanide and actinide doped glasses is not unexpected for bulk glass samples, the behavior of powders dispersed in index matching epoxy resins has, until recently, not been investigated. In figure 4, the fluorescence emission spectra of europium and uranium doped gel-glass dispersed in epoxy resin is presented. The peak at ~440 nm is due to the resin matrix. The peak at 520 nm is due to the hexavalent uranium. The large narrow peak at 614 nm, and the two "satellite" peaks adjacent, are from the trivalent europium. Recognizing that the fluorescence due to the resin may change over time, we sought to investigate whether this might effect the intensity of the inorganic fluorescence. In figure 5, the fluorescence intensity samarium and uranium doped gel-glass in epoxy is plotted as a function of irradiation at 365 nm. The fluorescence due to the epoxy resin (439 nm) declines dramatically as a function of UV irradiation, while the samarium fluorescence (602 nm) remains constant. This experiment has been repeated for a number of different combinations of gel-glass and resin, all yielding the same result. The inorganic fluorescence is highly stable and reliable.

A PRODUCT

In biological and medical science, well plate reading machines have become an indespensible laboratory tool in both research as well as in medical diagnostics. Plate readers are utilized for both ELISA test, such as HIV antibody test, as well as for PCR measurements[13]. Most measurements rely upon organic species, such as antibody conjugated organic dyes and proteins. Callibration and alignment is routinely required to assure measurement reliability. In figure 6, a photograph of a fluorescent wellplate reference block, incorporating lanthanide and actinide doped gel-glass in an organic polymer matrix is shown. This product is now commercially available. Typical uniformity of fluorescence intensity from well to well varies no more than 1.7 % over the entire 96 wells per plate. The only standard accepted by N.I.S.T., based on a solution of quinine sulfate, has a minimum well to well variation of over 3 percent or larger. More over, it must be freshly prepared for each measurement and has no shelf life after preparation.

ACKNOWLEDGEMENTS

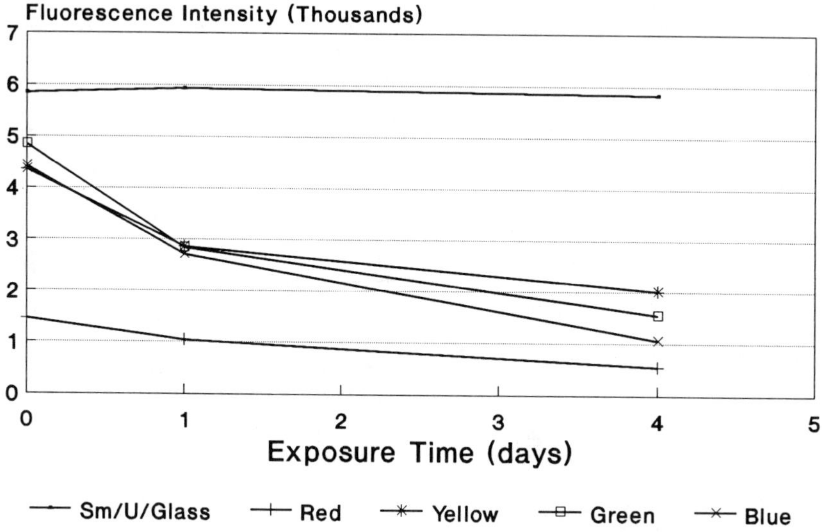

Figure 3: Fluorescence intensity of Sm^{3+}/U^{6+} doped gel-glass and four dye doped polymer samples as a function of 254 nm UV irradiation.

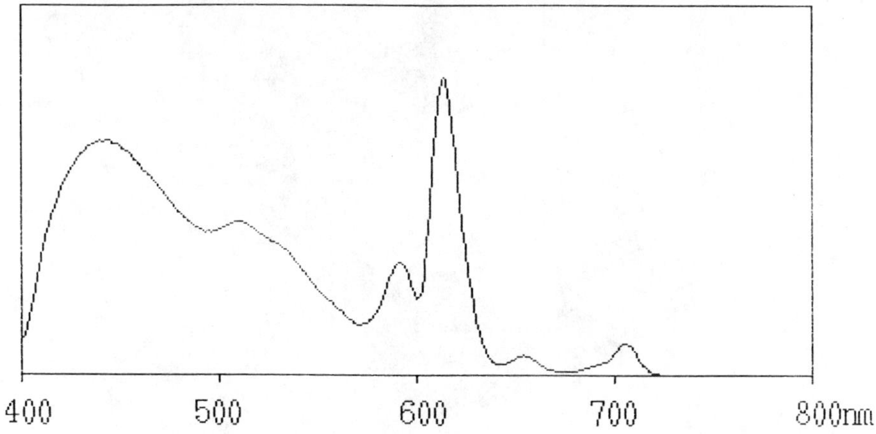

Figure 4: Fluorescence spectra of Eu^{3+}/U^{6+} doped gel-glass in epoxy resin matrix.

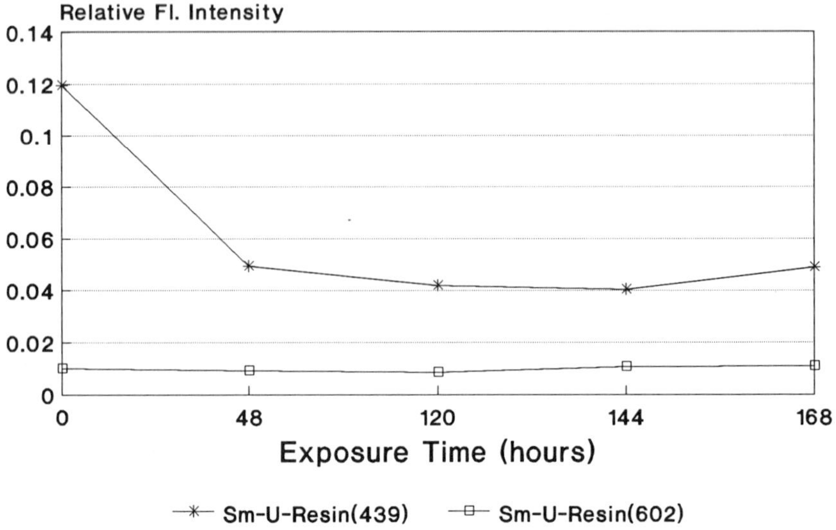

Figure 5: Fluorescence intensity of Sm^{3+} (602nm) and epoxy resin (439nm) as a function of 365 nm UV irradiation.

Figure 6: Fluorescence wellplate reader reference block.

The assistance of Mr. Alex Almazan in sample preparation and Mrs. Barbara M. Pope in manuscript preparation is appreciated.

REFERENCES

1. W. A. Weyl, "Coloured Glasses" (Society of Glass Technology, Sheffield, 1951).
2. I. Fanderlik, "Silica Glass and its Application" (Elsevier, Amsterdam, 1990).
3. D. R. Uhlmann and N. J. Kreidl, eds.,"Optical Properties of Glass" (American Ceramic Society, Westerville, 1991).
4. B. I. Lee and E. J. A. Pope, eds.,"Chemical Processing of Ceramics", (Marcel-Dekker, New York, 1994).
5. E. J. A. Pope, S. Sakka, and L. C. Klein, eds., "Sol-Gel Science and Technology" (American Ceramic Society, Westerville, 1995).
6. E. J. A. Pope, "Sol-gel Optical Nanocomposites", J. Sol-Gel Sci. & Tech., 2(1994) 717-722.
7. T. M. Che, et al, "Electro-Optic DC Kerr Effects and Solid-State Deuteron NMR Studies of Stable Gel-derived Glass-Organic Polymer Composites, J. Non-Cryst. Sol.,102(1988)280-287.
8. E. J. A. Pope, "Optically-active sol-gel microspheres for flat-panel color displays", in Sol-Gel Optics III, J. D. Mackenzie, ed (S.P.I.E., Bellingham, 1994) 536-545.
9. E. J. A. Pope, "Fiber-optic microsensors employing optically-active silica microspheres", in Advances in Fluorescence Sensing Technology II, J. R. Lakowicz, ed. (S.P.I.E., Bellingham, 1995) 245-256.
10. E. J. A. Pope, "Photochromic Nanocomposites", in Sol-Gel Optics III, J. D. Mackenzie, ed. (S.P.I.E., Bellingham, 1994) pp. 410-421.
11. F. J. Duarte and L. W. Hillman, eds., "Dye Laser Principles" (Academic Press, Boston, 1990).
12. E. J. A. Pope and J. D. Mackenzie, "Incorporation of Organic Dyes in Polymer/Oxide Composites", Mat. Res. Soc. Bull., 12[3](1987)pp. 29-31.
13. L. He, "The Efficient Use of PCR in Multiwell Plates in Mental Illness Linkage Analysis", Application Note,

PHOTOSENSITIVITY IN A SILICA-BASED SOL-GEL GLASS

D.M. Korwin* and L.D. Pye
Institute of Glass Science and Engineering
NYSCC at Alfred University, Alfred, NY 14802 USA

INTRODUCTION

Photosensitivity in glasses containing Au and Ce was first reported by Dalton in 1943 [1] and later reaffirmed by Stookey [2]. The photothermal reduction of Au ions to form metallic colloids was determined to be responsible for the "ruby" color produced in these glasses. In this work, the photosensitive effect has been confirmed for the first time in a silica sol-gel glass containing Au and Ce. Two methods of producing Au colloids in this glass were investigated, one involved a short ultraviolet (UV) exposure followed by a thermal treatment, the other a simultaneous UV irradiation and thermal treatment. Colloid formation was studied using optical absorption spectroscopy, whereas the role of Ce in the photosensitive process was elucidated using optical absorption and fluorescence spectroscopic techniques.

EXPERIMENTAL PROCEDURE

Gold- and Ce-doped silica glasses were prepared from acid-catalyzed sols using the alkoxide tetraethylorthosilicate (TEOS) as a source for SiO_2. The reaction conditions for the sol were $[H_2O]/[OC_2H_5]$ = 1.0 (molar) and the ethanol/alkoxide ratio was 1.0 (by volume). Gold was added in a concentration of 0.025 weight percent as tetrachloroauric acid ($HAuCl_4 \cdot 3H_2O$). The concentration of cerium was also 0.025%, with $Ce(NO_3)_3 \cdot 6H_2O$ as the source. The system was hydrolyzed using 0.04 N HCl as the source of water. After hydrolysis, the samples were thermally aged and slowly dried to a rigid state. Samples were then exposed to UV radiation from a 1000 W Hg-Xe lamp. Gold colloids were produced in the samples by two different procedures. In the first method, specimens were irradiated for 3 min, then separated into 4 groups for thermal treatments at 30, 40, 50 and 60°C. The second method employed continuous UV irradiation of the sample at a constant temperature of 30°C. UV-visible absorption spectroscopy was performed on 3 mm thick samples from both groups using a Beckman model DU-70 spectrophotometer. The fluorescence spectra of 6 to 10 mm thick samples irradiated for 0 min, 20 min and 80 min, were measured using a Spex fluorolog spectrometer. Excitation scans were made while monitoring the 361 nm emission wavelength.

RESULTS AND DISCUSSION

The role of Ce in the photosensitive process may be elucidated from the results of optical absorption and fluorescence measurements. Figure 1 shows the UV-visible absorption spectra of a sol-gel glass containing 0.025% Au and 0.025% Ce as a function of UV irradiation time ranging from no exposure to 180 seconds. The absorption band at 320 nm is assigned to Ce^{3+}, consistent with Ce^{3+} absorption in a variety of silicate glasses [3]. In the presence of strong UV radiation (E=hv) in the range of 300 to 350 nm, Ce^{3+} is photoionized according to the following equation:

$$Ce^{3+} + h\nu \leftrightarrow Ce^{4+} + e^- \quad (1)$$

Figure 1 shows that the Ce^{3+} concentration decreases with increasing UV exposure time as indicated by eqn. (1). According to these results, no Ce^{3+} is detected after 180 s of UV exposure. The presence of Ce^{3+} can also be revealed by fluorescence measurements as is shown in figure 2. Fluorescence of Ce^{3+} was monitored by the intensity of emitted light at 361 nm. This emission wavelength is in agreement with that observed for Ce^{3+} in silicate glasses [4]. The decrease of emitted intensity at 361 nm with increasing exposure indicates that Ce^{3+} has been oxidized to optically inactive Ce^{4+}.

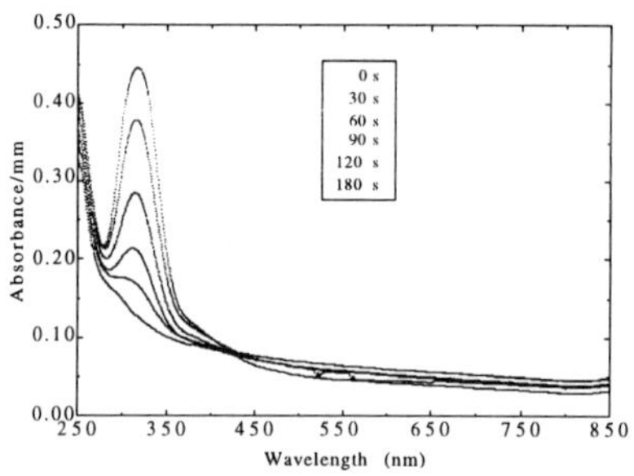

Figure 1. Optical absorption spectra for a sol-gel glass containing Au and Ce as a function of UV irradiation time. Irradiation times in the legend correspond to consecutively lower spectra in the figure.

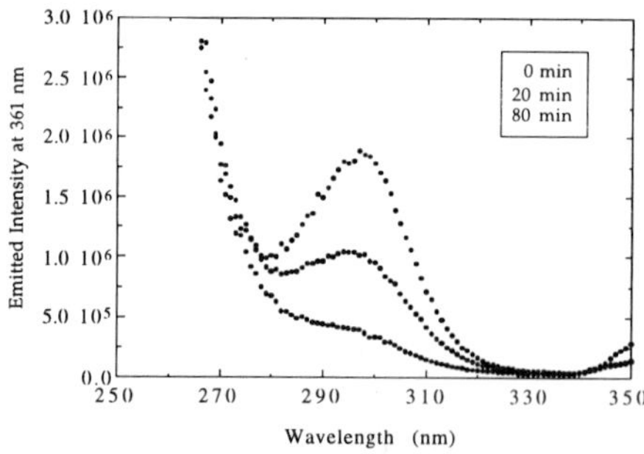

Figure 2. Fluorescence excitation spectra for a sol-gel glass containing Au and Ce as a function of UV irradiation time. Irradiation times in the legend correspond to consecutively lower spectra in the figure.

The data of figure 2 also indicate that Ce^{3+} ions persist until some time between 20 and 80 min of UV irradiation, whereas the optical absorption did not show the presence of this ion. Differences between the absorption and fluorescence data may be due to the sensitivity of the two techniques, as well as the size of the sample employed in each measurement.

Figure 3 shows the UV-visible absorption spectra of a sol-gel glass irradiated for 3 min, followed by thermal treatment at 50°C. The classical absorption band for colloidal Au is observed to develop at a wavelength of 545 nm [5], increasing in intensity as the heat-treatment time increases. The shift of the band to a wavelength of 550 nm at 80 min is attributable to an increase in size of the Au colloids. Similar spectra were recorded for the glasses heat treated at 40 and 60°C. It is noted in all of these spectra that the Ce^{3+} absorption band alluded to previously begins to reemerge with increasing time.

To compare the samples heat treated at different temperatures, the volume fraction of Au precipitated was calculated following the method of Turkalo and Doremus [6]. In this method, the volume fraction of Au precipitated, W, is assumed to be proportional to the ratio of absorbance at 436 nm at any time, to the final absorbance when colloid growth is assumed to be complete. The calculated volume fractions of Au precipitated for a fixed UV exposure of 3 min followed by heat treatments at 40, 50 and 60°C are presented in figure 4. The rate at which the Au precipitates increases with increasing temperature and is attributable to the increased kinetic energy available for diffusion of Au atoms.

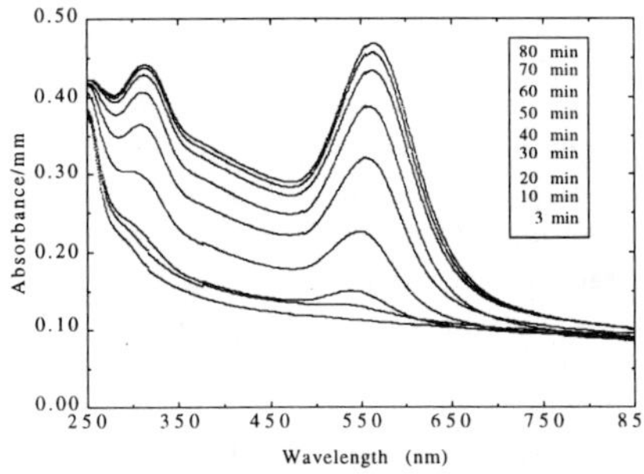

Figure 3. Optical absorption spectra for a silica sol-gel glass containing Au and Ce as a function of thermal treatment time at 50°C. Thermal treatment occurred after 3 min of UV exposure. Irradiation times in the legend correspond to consecutively lower spectra in the figure.

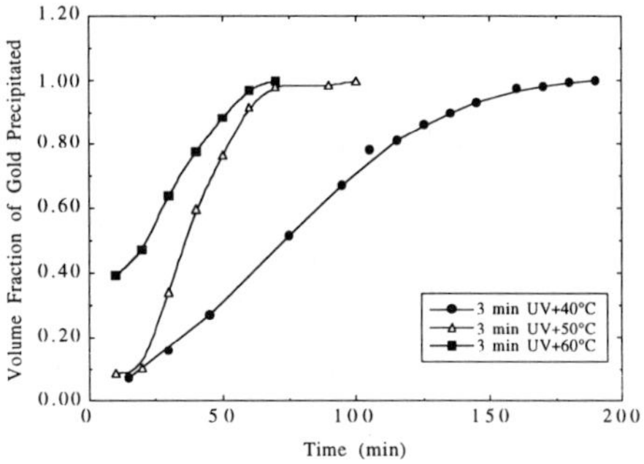

Figure 4. Volume fraction of Au precipitated as a function of thermal treatment time at 40°C, 50°C and 60°C. Thermal treatments follow a 3 min UV exposure.

The second procedure used to precipitate Au colloids in the silica matrix involved a thermal treatment at 30°C with simultaneous UV irradiation. Figure 5 shows the UV-visible absorption spectrum of a sol-gel glass irradiated with UV light for increasing amounts of time from 3 to 70 min at a temperature of 30°C. Again, the Au colloid absorption band is observed at a wavelength of about 545 nm, increasing in intensity and shifting to slightly longer wavelengths as the irradiation time increases. The comparison between thermal treatment at 30°C with and without simultaneous UV radiation is shown in figure 6. At 30°C with simultaneous UV irradiation, the sol-gel glass is photosensitive, resulting in Au colloid formation. Whereas at 30°C, without simultaneous UV irradiation, Au colloids do not form. From these results it is concluded that simultaneous UV exposure and heat treatment promote colloid formation.

The enhancement of Au colloid formation in the presence of UV radiation may be explained by considering the probability of reduction of the Au ions by electrons. The electron produced by the reaction in eqn. (1) is trapped in a state near the Ce^{4+} until sufficient thermal energy exists for the transfer of the electron to Au^{+} resulting in reduction to metallic Au^{0}[7]:

$$Au^{+} + e^{-} \leftrightarrow Au^{0} \tag{2}$$

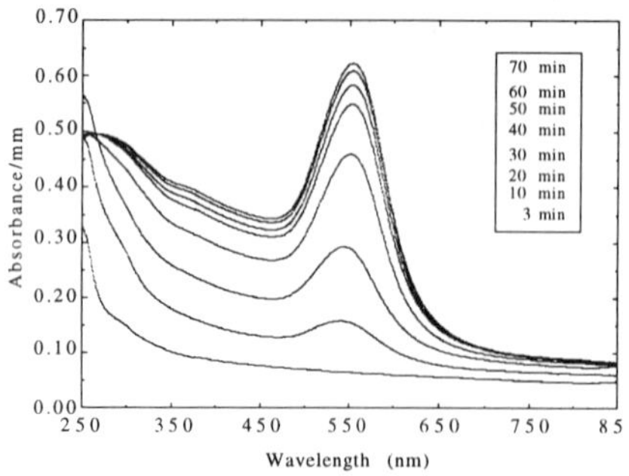

Figure 5. Optical absorption spectra for a silica sol-gel glass containing Au and Ce as a function of UV irradiation time at 30°C. Irradiation times in the legend correspond to consecutively lower spectra in the figure.

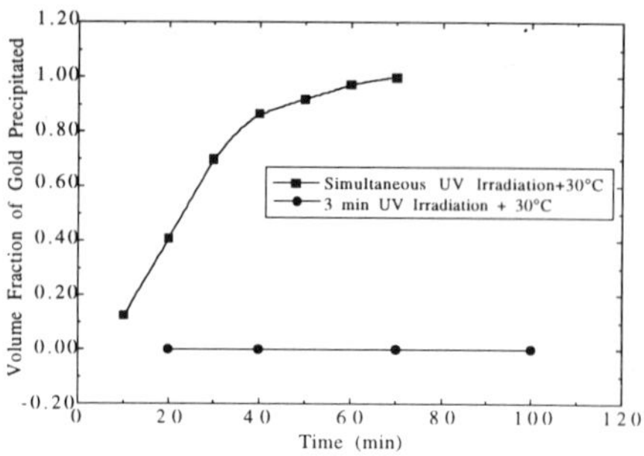

Figure 6. Volume fraction of Au precipitated as a function of treatment time at 30°C for sol-gel glasses treated with (a). continuous UV radiation and (b). no UV radiation during thermal step.

These gold atoms then coalesce by diffusion to form colloidal particles. Under the influence of UV radiation, eqn. (1) is forced to the right by LeChatelier's principle, ensuring a higher availability of electrons to reduce Au ions. In the absence of UV radiation, the Ce^{4+} ion may reabsorb the electron produced in eqn. (1) thereby reversing the reaction and lowering the availability of electrons. This hypothesis is supported by the data of figure 3 which clearly show that the Ce^{3+} absorption peak at 320 nm begins to return after 30 min of thermal treatment in the absence of simultaneous UV irradiation. Whereas, according to the absorption spectra of figure 5, the Ce^{3+} peak does not reemerge during simultaneous UV irradiation at 30°C. To our knowledge, this is the first time that photoassisted colloid formation has been observed in a sol-gel based material.

REFERENCES

1. R.H. Dalton, "Glass Article and Method of Making It", U.S. Pat. 2,326,012, 1943.
2. S.D. Stookey, "Colorization of Glass by Gold, Silver and Copper", *J. Am. Ceram. Soc.*, **32**[8] 246-249 (1949).
3. A. Paul, M. Mulholland and M.S. Zaman, "Ultraviolet Absorption of Ce (III) and Cerium (IV) in Some Simple Glasses", *J. Mater. Sci.*, **11** 2082-2086 (1976).
4. M.J. Weber, "Cerium-Activated Crystal and Glass Scintillators", pp. 99-124 in **Heavy Scintillators for Scientific and Industrial Applications**, Editions Frontieres, France, 1993.
5. R.H. Doremus, "Optical Properties of Small Gold Particles", *J. Chem. Phys.*, **40**[8] 2389-2396 (1964).
6. A.M. Turkalo and R.H. Doremus, "Electron Microscopy and Optical Properties of Small Gold and Silver Particles in Glass", *J. Mater. Sci.*, **11**[5] 903-907 (1976).
7. N.F. Borelli and T.P. Seward, III, *Photosensitive Glasses and Glass-Ceramics*, pp. 439-444 in **Volume 4 of the Engineered Materials Handbook: Ceramics and Glasses**, ASM International, 1991.

KEYWORD AND AUTHOR INDEX